BACK TO THE FARM

Jackie Spilker

On Target Enterprises
Matheson, CO 80830

CONTENTS

PREFACE

BACK TO THE FARM is written by Jackie Spilker, who left her suburban life and corporate career to pursue a farm career. Single with 2 children and no background in agriculture, she applied for and received financial aid from the Farm Service Agency, a division of the United State Department Of Agriculture to start her Eastern Colorado Ranch.

Jackie is 1 of only 10,400 women farm operators in her age category in the nation. She is also 1 of only handful of single women, her age that have started, own and operate an agricultural enterprise.

The information complied in this book is from the authors first hand experience. The author has taken the concept of this book from idea to reality. Not an easy task, being a woman in one of the most male dominated career fields in the nation.

About This Book

Millions of Americans continue to idealize a farming or ranching lifestyle. A simple country life, roots, living off the land, and a sense of community. All values and ideas that to this day continue to be a part of the All-American Dream.

America was born on the farm. Great men of history such as George Washington and Thomas Jefferson were both first farmers. When the United States was founded 90% of the population were farmers and ranchers. As recently as the 1950's, 17% of the nation's population were still living and working on their own farms and ranches. A large percentage of today's adult population can remember spending summer vacations on grandma and grandpas farm. And what they remember was a much simpler time. To their minds the grandparents did not have as much pressure, no corporate ladder, traffic jams, smog days. The never had to lock the front door and enjoyed close family, neighbors and a community that cared about one another.

Today, there are vast numbers of people in the United States that want to farm. And on top of that an even greater number that simply want to return to rural life.
The statistic and trends state the undeniable facts:

- 62% or 48 million Americans stated that their desire to move to the country is increasing, according to a recent Gallup Poll.

- There are 3 million new rural residents with the pace of rural growth accelerating.

- July of 1995, American Demographics reports that factors such as long-term economic changes favor rural areas, along with the belief by many Americans that small town life is better.

- All of the big cities, except New York City, are losing population, with ¾ of all rural counties gaining.
- June of 1996 the USDA reported that employment in non-metro counties is growing faster than metro.
- In a recent Gallup Poll 22% of Americans said they desired to live on a farm.

More and more individuals are living 20 to 50 miles from where they are employed to enjoy more enriching lifestyles. Additionally the new home office trend of is encouraging even more families to seek small towns and rural locations. People want to live in small towns and rural communities. Most simply need to know there are opportunities. Although 62% of Americans desire to move to the country only 34% act on that dream, why do the other 28% remain?

Although there is the desire on the part of Americans to get back to the country, these people have no idea how to follow their dreams...with the major drawback being financial considerations. Fear of not being able to make a living in a small town atmosphere keeps many city dwellers from making a break for the country. What many aspiring country residents don't know is that there are growing opportunities in Rural America. Rural business incentives are attracting large companies to small towns, cottage industries can be started and run more cost effectively from the country and individuals can derive a good income full or part time from the very ground they live on by farming or ranching on a small scale.

For individuals considering a farming or ranching lifestyle the fear maybe even greater. How to get started financially is a major stumbling block. This coupled with the general lack of knowledge of rural life is enough to thwart any dreams of playing out Lisa and Oliver on their own Green Acres. Young people have lost their families farming heritage and therefore the skills and knowledge required for farming. If an individual has the desire to farm, where do they find the education? Farming 101 is not taught in your major suburban high schools or colleges along with the rest of the sciences and arts. It is certainly not represented on career day. Agriculture is by many considered a lost art handed down through generations.

For those with a strong yearning to dabble in the oldest profession in history, terms such as crop rotation, commodities markets and animal husbandry are about as foreign to them as trying to understand Chinese. However, farming and ranching are not hard or mysterious. They are just largely un-taught. For the most part it can be thought of as a factory without walls. To simplify...products, supply, demand, overhead, buy low – sell high. The big difference - some of your products may cluck or moo.

Today, many new farmers are proving you don't need a farming background or have a family farm handed down, to succeed in farming. New methods of farming plus financial incentives are making it easier than anytime in the recent history to get started on one's own farm.

- New farming products and technologies are producing more profits from fewer acres. Fewer acres and fewer inputs require less cost in initial startup.
- With Internet access to college, state and federal agricultural departments and local extension agents farming knowledge is only a keyboard and modem

away. The Department of Agriculture, Extension Offices and many government agencies are set up to give free expert advise on the what, when and how of agriculture in any area of the country. Internet accesses to agricultural libraries, farming and ranching publications, breeders and equipment dealers can lend a wealth of information to those considering an agricultural enterprise.

- State and Federal financial incentives are aiding new farmers with low interest loans and grants to get started in a farming future.

Why farm programs and agricultural incentives? According to the USDA "Since the late 1920's, American farm policy has tried to encourage the production of adequate supplies of food and fiber and to maintain reasonable prices for consumers while, at the same time, assuring farmers a fair return on investment". Information on the beginning farmer and rancher programs, referenced in this book, can be difficult to find. The USDA or participating states do not take out large print or TV ad's proclaiming free or low interest money for those who want to farm. Ironically the group that is made most aware of these programs is the group that is already farming. The same holds true for programs intended to spur rural growth such as money for rural housing.

Not only are early retirees and "cashing outers" moving to the country but the federal government is also attempting to encourage low to moderate-income families to come back to rural areas. State and Federal agencies are offering:

- No down payment mortgages for rural low to moderate-income families.

- Grant and low interest loans to update and remodel rural housing.

- Rural rental assistance.

These programs and that fact that housing prices in some rural areas are less than half that of comparable city houses are enabling individuals to live in areas they never thought possible for them and their families.

Within the pages of this book you will find much of the information and knowledge required to make your thoughts of country life or perhaps deriving an income off the land seem like a realistic idea. It is intended to be a no frills reference book and decision guide. Covering economic and social considerations, fact, figures and examples related to rural employment opportunities, housing and starting an agricultural enterprise.

Unlike other rural or country living books it does not elaborately detail the lives of individuals that gave up tremendous careers and incomes to take an early retirement in the country. It is geared toward "plain folk" in search of some peace and quiet. After all everyone deserves clean water, air and to be safe in their homes and neighborhoods. The basic premise of Back To The Farm is that anyone can move to a small community or rural area and live a satisfying, safe and secure life.

Chapter 1 America Going Home - A Quiet Revolution

"I don't like the city better, the more I see it, but worse, I am ashamed of my eyes that behold it. It is a thousand times meaner than I could have imagined" Henry David Thoreau

"The Boonies Are Booming!" according to a recent Business Week Report. There is a quiet revolution going on in America today. The return to Rural America is one of the biggest trend setting stories of the 1990's. For the first time in 200 years more people are moving to the country than to the city.
Between 1990 and 1994, 3 in 4 rural counties gained in population. This is in direct contrast to demise of rural America that played out in the 1980's. From 1990 to 1991, out of the 2,304 rural counties, more than 67% gained in population. Most rural communities in the United States are now growing at their fastest rate in more than 20 years.

In "The Rural Rebound: Recent Non-metropolitan Demographic Trends in the United States". Kenneth Johnson from Loyola University and Calvin Beale with the USDA stated "Since 1980, non-metropolitan population growth rates have rebounded from the low levels of the 1980's. Population growth is now widespread geographically and is occurring in counties with a variety of economic specialties. In July of 1996, non-metropolitan areas of the United States contained 53.8 million residents, a gain of nearly 3.0 million since April of 1990. This gain is already more than twice as great as that during the entire decade of the 1980's. Most of the growth came from net migration rather than from the natural increase (births-deaths) that has traditionally fueled non-metropolitan growth."

Over a four year time period, from 1990 to 1994, in states like Iowa, 60% of the towns grew to contrast to the 8 out of 10 that experienced shrinkage in the 1980's. In addition, a recent study by the University Of Iowa showed more than half the merchants in 30 communities had planned on expanding.

Why is this phenomenon taking place? This new country craze seems to have its roots in social issues. The greed of the 80's and the high price of materialism have left people feeling empty, lonely and disconnected. This is evident in the huge number of Americans being treated for depression and anxiety.

Substance abuse is at an all time high. And masses of lonely people habitually using chat rooms on the Internet are evidence that people are yearning to connect with someone. Princess Di, in one of her last interviews, stated she felt the greatest need in the world was for the masses of people that felt unloved.

The search for more meaning in life is changing social attitudes. Earning money to keep up with the neighbors is becoming an antiquated value. Families in debt up to their ears working over 60 hours a week and not spending anytime with each other is causing suburban and urban overload. The sense of life controlling you not you controlling your life is causing feelings of suspicion, hostility and aggression. America's preoccupation with material success is causing burnout. Many studies have shown that the high incidents of stress related diseases and mental dysfunction are closely related to high levels of stress caused by urban life. Depression is now the third most common disease in America.

"This movement is pervasive", according to sociologist Stephen Warner of the University of Illinois at Chicago. "This is not something simply happening to the burnout's from Wall Street. There is an American phenomenon going on that crosses all social lines". Unplugging from urban and suburban life is a trend of the future.

What are the reasons for the behind what is touted as the "Rural Rebound Movement"? Factors driving the rural rebound movement are a preference to small town or rural life, the demise of urban areas, metro recession fueled by corporate restructuring, the increase of early retirements to lower cost of living areas, and the new technologies allowing more employment from non metro areas.

A recent study by the Roper Organization found a strong longing on the part of Americans to live in a small town or rural area. When asked where they would like to live right now, 33% said their first choice would be a small town or rural area.

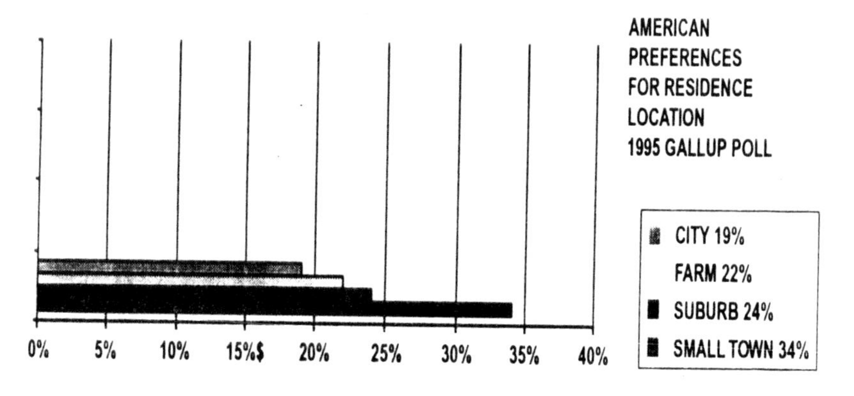

As the troubles in Urban America such as the expense, crime, pollution, quality of schools worsen year to year it is creating an exodus back to rural areas. Half of all urban U.S. families can no longer afford the American Dream.

The cost of housing, utilities, transportation in the nation's metro areas has turned life into the American Nightmare. The middle-class lifestyle in a metropolitan area is out of reach for the majority of city inhabitants. 38 million Americans live in poverty. Conditions in the inner cities are comparable with the Great Depression. 10% of Americans control all the wealth creating a tremendous gap between rich and poor. The inner city and most metro areas have completely deteriorated. In a 1990 Gallup Poll half of the families surveyed stated they desired to move to a community of 10,000 or less. "Some urban experts believe that the major US City is a thing of the past". The American Council of Life Insurance.

The corporate restructuring of the 1990's has fueled to some part the rural rebound movement. Tremendous emphasis placed on the bottom line caused many U.S. Corporations to cutback creating millions of displaced workers.

These workers, many of whom feel used and then dumped by those companies have left with large severance packages. In addition most have large equities in suburban or urban homes, which has placed them in the idea situation to go rural. Being able to buy twice the house for half the money. Their distrust in corporate America and current financial position has caused many of these boardroom burnout's to take early retirement in the country.

Urban Vs. Rural

The longing to move to the country for most is not enough. Financial and social considerations play a major role in the decision to move or to stay. There are many families that are not in the financial position to cash out and take an early retirement to the country or are not fortunate enough to own a business that can be run from a rural area.

The greater portion of the population that desires to move to the country are not of such financial means. Can they afford to move? Or can they not afford to move? It is certain the deteriorating circumstances in metro areas are not getting any better. But urban and suburban surroundings, good, bad or indifferent are familiar and moving to and living in the country is unfamiliar. Nevertheless, sometimes the fear of the unknown can simply be overcome by studying an issue until a level of comfort with the subject is met. Moving to the country may not be for everyone. There are good and bad reasons to stay or go. Plain logic and reliable facts can be used to draw conclusions and sway opinion one way or another. The issue for most families may possibly just need some investigation. Conclusions to base a decision upon.

Income

I have chosen to start urban to rural comparisons with income since undoubtedly seems to be the largest obstacle for most people that desire to move to the country. One must remember that all things are relative. If a $250,000 home in California would only cost you $60,000 in Iowa you can afford to make less.

The following salary comparison is of randomly chosen urban and rural cities.

The comparison was based on an urban salary of $50,000 a year.
In other words if you make $50,000 In New York City, you only have to make $11,683 to enjoy the same standard of living in Enid, Oklahoma, (i.e. live in the same size house, pay the same for food, etc.).

Most urbanites looking to make the switch to rural life often forget that the cost of living in rural areas can be less than half that of the city. In most cases, families can purchase larger homes on more property for less money.

Many city folks are horrified at the salaries in small towns or rural areas. Good jobs in rural area's may pay $8.00 an hour but if you only have to pay $35,000 for a comparable 3 bedroom home in a safer neighborhood with better schools and clean air and water, financial considerations may make you contemplate a move sooner than expected.

Based on an Urban Salary of $50,000

	PACATELLO, ID	LARMIE, WY	ARKANSAS CITY, KS	ENID, OK
LOS ANGELES, CA	$24,253	$24,178	$23,745	$23,086
NEW YORK, NY	$12,274	$12,235	$12,016	$11,683
WASHINGTON, DC	$28,747	$28,658	$28,145	$27,364
BOSTON, MA	$24,361	$24,286	$23,851	$23,189
CHICAGO, IL	$26,067	$25,986	$25,521	$24,813

Source: The Salary Calculator - Homebuyer's Fair, Center For Mobility Resources www.homefair.com/home/?NETSCAPE_LIVEWIRE.scrs

Crime

Public Opinion Magazine recently reported that 59% of Americans feel it is too risky to go for a walk in their neighborhood after dark.

Violent crime such as rape, murder and robbery add up to 10% of crimes being committed in America. With metropolitan cities making up 28% more crime than in small towns. Crime is the number one concern of all Americans. America has the highest rate of incarceration of all the industrialized nations. It is now twice what it was 20 years ago.

In the 21st Century our greatest threat wont be from a foreign enemy, but from ourselves. The annual cost to this country of violent crime is over 4 billion with 42 million crimes being committed including 11 million violent crimes. Juvenile crime is up by 75% with the juvenile murder rate up by 165%.

Crime is one of the major considerations for most rural rebounders. They simply want to distance themselves and their families from this element. The search for safety seems to be leading people back to the country. Crime does occur in rural areas but with far less frequency and violence as in the city.

CITY	ROBBERIES	RAPES	HOMICIDES	AGGRAVATED ASSAULTS	MOTOR VEHICLE THEFTS	CRIME LAB INDEX
CHICAGO, IL	1111	107	30	1448	1337	260
NEW YORK CITY, NY	1148	35	21	898	857	185
WASHINGTON, DC	1428	21	66	1197	1878	331
LARMIE, WY	4	34	0	161	127	21
POCATELLO, ID	24	18	2	73	219	19
ARKANSAS CITY, KS	47	47	0	345	266	38

SOURCE: THE RELOCATION CRIME LAB - HOMEBUYERS FAIR
THE CENTER FOR MOBILITY RESOURCES. www.homefair.com/home/?NETSCAPE_LIVEWIRE.scrs

Health

A recent survey conducted by Parade Magazine found we could be taking better care of ourselves. "Americans are living longer, but they aren't living healthier". Says Dr. Claude Lenfant, director of the National Heart, Lung and Blood Institute in Bethesda Md.

And "Living in a large city shortens life expectancy" states Norman Shealy, founder of the Shealy Institute for Comprehensive Care and Pain Management in Springfield Missouri.

The plain truth of the matter is that the city can kill you.

- Incidence of cancer is 6% higher in urban areas.
- Air pollution kills over 80,000 Americans annually (more than killed in traffic accidents). With 94 cities violating the Clean Air Act. The air in 60 metropolitan areas reach unsafe levels several times a year with Californians breathing unhealthy air 232 days of the year.
- Americans landfills are expected to annually average 160 millions tons of trash that are leaching into the ground water.

Urban stresses are taking their toll. Hypertension, stroke, cancer, heart disease, pneumonia and ulcers are among the major stress related killers.

"Thirty-five years of extra life. That may sound exaggerated. But it is the actual statistical difference between the life expectancy of a 37 year old, overweight, divorced male who lives in New York City, smokes 2 packs of cigarettes, and takes several alcoholic drinks daily, performs no exercise, and is employed at a dull, routine office job and a slender, happily married Norwegian male, also aged 37, who abstains fror all forms of substance abuse, and who leads an active outdoor life working on his own farm." Norman D. Ford - The 50 Healthiest Place to Live and Retire in the United States.

Schools

The nations metro school system is plagued with low academic scores, overcrowded and under supervised classes, drug dealers and crime. This has caused worried parents to look elsewhere to educate their children in a safe environment. There are 3 alternatives to metro schools - selective choice in a rural school, private school and home schooling. The number of home schooled kids has jumped to over 1 million and 1 out of 9 children attend private schools.

Rural Schools can be quiet an adjustment for metro schooled kids. Strict discipline in the classroom and more parents volunteering equates to expectations of good behavior. In small communities the parents of children with bad behavior are not subject to anonymity as in metro areas. Rural schools tend to not graduate children undeserving of their diploma. The pluses of small town schools include: high visibility and recognition for children that have excelled in school academics, sports, music, etc., greater participation of students in every activity (every body is needed to make up a team), smaller classes, higher academic standards.

The downside is of course monetary. Not enough money sometimes means a cutback in sports equipment, field trips, musical instruments, computer equipment and theater. After school child care seems to be easier to find due to the number of stay at home mothers and most rural kids, especially farm kids, have no time to loiter after school, they have chores to do. A high number of small town and rural kias are also involved in after school sports and 4-H.

SCHOOL	SCHOOL POPULATION	RATIO STUDENTS TO TEACHERS	ACT SCORES	% OF SENIORS GOING TO COLLEGE
DALLAS CO., TX	10,464	21	14.2	42
SAN DIEGO CO., CA	5,211	22	17.7	86
LARMIE CO., WY	445	7	23.5	55
TWIN FALLS CO., ID	338	9	33.8	80

SOURCE: NSRS

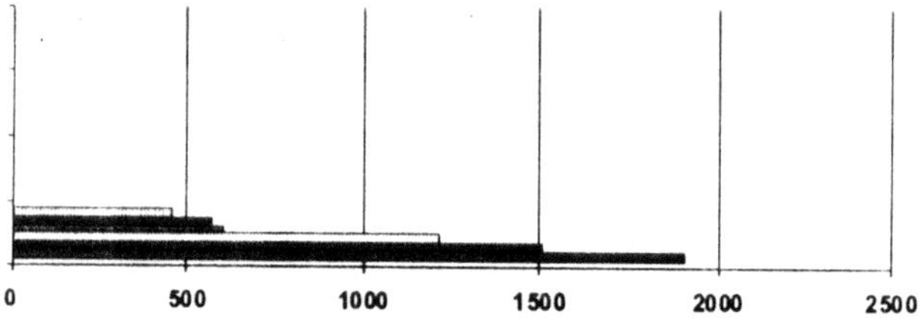

Property Tax Comparison
1995 - $80,000 Home
Wyoming 450
Oklahoma 580
Idaho 600
Washington 1200
Illinois 1500
New York 1900

0 500 1000 1500 2000 2500

Taxes And Services

The level of taxes generally supports the level of service. Lower taxes support minimal services. Services such as law enforcement, fire protection and zoning are all affected by the amount of revenue generated by taxes.

Law Enforcement may consist of a small Sheriff's Office that covers a large district and is sometimes stretched thin. Fire protection is very minimal if not non-existent in rural areas. Most rural fire departments are staffed with volunteers. When shopping for rural and small town homes and properties it is imperative to check with your insurance company. When I purchased my ranch I was unaware that my fire code rating was one of the worst in the nation. That calculated into a higher insurance premium. If your not careful your insurance rates in a rural area can knock you out of the ballpark when qualifying for your mortgage.

Financial Pro's and Con's of Rural Living

Expect to pay more for:

Utilities	Electric and Propane for the most part tend to be higher in rural areas. Most rural electric companies are generally co-ops and tend to pay more and pass the cost on to the rural consumer. Propane prices can be low in the summer and very high in the winter. It's the old supply and demand theory. You can cut your propane heating bill sometimes almost in half by having your tank filled in the summer when the prices are low.
Vehicle Maintenance	Dirt roads are hard on the suspension system, tires, paint and windshield.
Appliances/Electronics	Less competition and higher delivery rates are passed onto the rural consumer. Your best bet is to make a trip into the big city for big ticket items.
New Cars	Same holds true for new cars. One exception maybe to buy the vehicle over the Internet from an auto broker. They

can generally do better than a big city lot. You may have to pay for delivery but maybe not if the truck is passing through your area.

Expect To Pay Less For:

Housing	Housing and property in small towns and rural areas can be at least half the price of comparable homes in large metro areas. Prices for raw land are even more affordable. I would doubt you could find land for $50.00 an acre today. However gentle rolling hilly eastern Colorado farmland is going for $250.00 an acre so you never know.
Rent	Charging what the market will bear. Low incomes equate to lower rents.
Services	Low Overhead. Lower wages and low rent mean services such as attorneys, hair stylists, mechanics, contractors generally cost the consumer less in small towns.
Clothing	A casual atmosphere means getting away with less expensive clothing.
Car Insurance	Since automobile insurance is rated by the number of thefts and accidents in any given area. Expect to pay considerably less here.
Entertainment	Picnics, hiking and enjoying the outdoors is free. Movie theaters in small town's charge less although the first run movie is shown a month later than the first showing in the big city. Small dish companies such as Direct TV charge about the same as metro cable companies.
Food	Local produce and locally produced meats are fresher and less expensive. Sometimes you can pay more for items due to transportation costs being passed on.
Dining Out	Restaurants in small towns don't do a booming business especially on weekdays. They tend to run daily specials to attract business on off days.
No Temptations	Expect to pay less living in the country due to the simply fact there is just no where to spend your money. I have gone for days without leaving the my place. If you live far from shopping. You tend to go in with a list and stick to it. It is amazing how much money you can save when the mall becomes such a hassle and inconvenience to get to.

Expect To Pay The Same For:

Existing Personal Bills	Credit Cards, Student Loans, Personal and Car Loans.

Profile Of A Small Town Newspaper - Limon Leader
Limon, Colorado

Population 2,400
Location: Eastern Colorado Plains

Homes Available In And Around Limon

Remodeled, 3 Bedroom, 1 Bath, 1 Car Garage	$65,000
2 Bedroom, 1 Bath, Owner Motivated	$28,000
2 Bedroom, 1 Bath, Full Basement, 3.2 Acres	$32,000
3 Bedroom, 2 Bath	$36,000
10 Unit Apartment Building (All Leased)	$125,000
Liquor Store with all inventory and equipment	$57,000
Restaurant and All Equipment	$50,000

For Rent

Mobile Home For Rent	$260/month
2 Bedroom Apartment	$300/month
Office Space	$375/month

Help Wanted

Arbys Manager, 4 nights a week, $17,274 per year
2 openings for Service Technicians with Chrysler or GM experience
McDonalds, $5.50 hour
Pizza Hut, All Positions
Chemical Dependence Therapist II, FT
Route Driver for Dairy
Full Time Social Worker
LPN in Rehab Center
Housekeeper, Econo Lodge
Clerk/Cashier wanted at Total
Business Education/Computer Instructor wanted at Big Sandy School District
School to Career Coordinator, Limon School District
Middle Math School Teacher, Genoa Hugo School
Register Nurse - Kit Carson County

Groceries

Tomato Sauce	6/$1.00
6 Pack Pepsi	$1.79
Red Delicious Apples	$3#/$1.00
Toino's Pizza	.99
Post Honeycomb	2/$4.00

Front Page News

Flood Victims can get FEMA assistance
Town Board changes August meeting date
Linc-Up to begin fall session
G-H cheerleader fund raiser at Fair
Limon PD gains additional officer
School Registration Set For Limon
SHARE - helping the community help themselves.
Band Members Uniform Information

Changes In Rural America

Not only are city dwellers packing up and heading for the country, a substantial majority of the 22% of the nations population that are rural residents are happy to stay right where they are.

In a recent poll by Communicating for Agriculture, 90% of rural Americans surveyed said they would tell their children that a rural lifestyle is "the good life". More than 2/3 of the respondents said they felt is was likely that their children will decide to settle down in rural areas. CA's president Wayne Nelson stated, "The survey confirms what many of us have felt all along - that rural America is a great place to live, work and raise a family. The fact is, people want to live in rural areas, buy they need to know there are opportunities".

People in small towns today have access to virtually everything that the city populations do. Computers, faxes, modems, Internet access, satellite TV, economical cars and a completed national highway system have now made country living less isolated and more economically practical. Social tensions of the past such as discrimination of ethnic, gender and sexual orientation have eased tremendously in small towns and rural areas Lending itself to more acceptance and ease of blending of all groups of people. Many young people now contemplating leaving the small hometown environment have taken a second look at staying or coming back after college.

With more people staying put and the growth in rural areas comes the reversal of small town deterioration. The rural demise of the 1980's that was caused by a prolonged rural recession, farm crisis and higher job growth rate in the cities appears to be over.

The rural rebound movement has brought about significant changes in Rural America. Many of these changes have been for the better. Improvements in utilities and a highway system that is completed have promoted a better standard of living in rural America. While poverty stricken rural America continues to make headlines, in the past decade, these problems have reversed themselves and that trend is continuing.

Today rural families are better housed and more likely to own their own homes. Only 2% of rural homes were considered substandard in 1990 (lacking complete plumbing) while 75% were considered substandard in the 1940's. High School completion rates are now nearing those found in urban schools. 75% of rural residents own their homes today instead of 50% in the 1940's. Less than 2% of households live in a home with fewer rooms than occupants, down from the 1940 figure of 25%.

The number of counties with the population under poverty level has made a tremendous reversal in the past 30 years. In 1990 the number of poverty counties were 765 in stark contrast to 2,083 in 1960.

However problems remain. 25% of the nations poor live in rural America. Lower wages and a lower college completion rate is still prevalent in these areas.

Vast differences in the rural experience can be seen from one part of the country to the other largely depending on the country type. County types like Service, Retirement, Federal Land and Manufacturing are demonstrating a higher standard of living while farming and mining counties are seeing a small to flat rise in their standard of living.

Manufacturing Counties: Home to 33% of the rural population and employs twice as many people as does farming. Jobs in manufacturing counties tend to pay more and

grew by 2.8% in the 90's.
Service Counties: The Service sector includes utilities, transportation, finance, insurance, real estate and others. Service Counties are the fastest growing. From 1979 to 1989 they comprised of over 3 million non-metro jobs and accounted for nearly 83% of all rural jobs. However jobs in service counties tended to pay slightly less than other rural counties.
Federal Land Counties: These are counties in which 30% or more of land is owned by the Federal Government. Job growth in these counties is strong and outpaced even metro based growth. Family income in these counties was nearly 8% higher than other rural counties.
Retirement-Destination Counties: Have experienced a 15% increase of people age 60 and older in the 1980's. Natural amenities including mountains, sea shores and mild climates draw retires and tourists to these counties. These counties had the highest rate of growth over the past decade. And had a 26% earnings growth and a 34% job rate growth. However earnings in these jobs are about 5% less that other rural counties.
Farming and Mining Counties: Farming counties are only 9% of the rural counties in direct contrast to the vast number of farming communities of the 1950's. Because of new advances in technology fewer people are needed in jobs in Farming and Mining Communities. For the most part the jobs that are left are paying well and fall above the average of other types of rural counties.

Chapter 2 - Opportunities In Rural America

The biggest concern for would-be country dwellers is how to make a living. This argument has always surprised me. It is just as easy to be unemployed in the city as in the country. About 400,000 people were laid off in 1996 from downsizing alone in addition to the 4.5 million that lost their jobs the preceding 10 years, according to the Kiplinger Washington Letter. Yes one can go on many more interviews in the city. However is their not more competition for the same job?

When I moved to the eastern plains of Colorado, I moved with no job and about $3,000 in savings. Just diving right in. After settling in I sent a flyer to neighboring small towns advertising my skills in marketing and office management. Since then I believe I have had more job offers living here than living in the city. What I did not realize is the shortage of qualified applicants in rural areas. What I have grown to understand from small town employers and the states job service office is that most high school students that left to go to college never returned, which has left a virtually un-skilled, under educated labor pool.

Due to rural rebound and the growth in small town America many service companies such as communications, specializing in products like cell phone service, are expanding into the country. These companies are_experiencing difficulties in finding qualified applicants and are for the most part relocating employees into these areas. Rural businesses saw notable gains in sales for the first half of the 90's. Wal Mart quickly became the nations largest retailer by expanding though out small town America over the past 20 years. The increase in rural population has caught the attention of the nation's big business. Many of who hired demographers to track this movement and plan for full blown expansion into these areas.

City jobs as a whole still pay more than rural jobs but rural income is on the rise. "In 15 states, rural counties rank highest in per-capita income." Business Week 10/95. With my move to the country I fully expected to make less. However I found my skills where in such high demand I have discovered I have made more money in the country than in the city. This is in addition to my farm income and my really low mortgage payment on a house, three large barns and property (10 times more than I could afford in the city). All of this plus I raise my own food and have become more frugal due to the logistics of being so far away from shopping. I do believe I am in a much better position financially than I ever was living in the Burbs. Even if I must drive 60 miles or more to get to good employment.

What is the difference between driving that far at 50 miles an hour down dirt roads, watching the sun come up over rolling fields of hay and windmills, than sitting in rush hour traffic for an hour to get 20 miles to work in smog next to an angry motorist sporting a gun in the glove box?

Employment Options

You have 3 employment options when living in the country. Find a large company, perhaps the one you are working for now, that will allow you to work from home and telecommute, work for an existing rural business or start your own business. Perhaps a combination of the three options will work best for you. Many individuals just starting out in their own business find it far less financially draining to work part time while building a business of their own.

Companies At Home In Rural America

According to the Bureau of Labor Statistics 31% of us now work from home. Times certainly have changed. 20 years ago imagine the furor it would have caused if you had marched into your bosses office and requested to work from your new home in the country over 200 miles away and only come into the office maybe once a week if not less. You would have been marching right to the un-employment line.

Now, this is a common occurrence in American Business. In 1994 the home workforce comprised of 36 million, with a 6.5% increase over the previous year. Companies like A T & T, who have more than 7,000 of it's workers at home, have found that the savings in real estate and office rental space alone is enormous. Companies such as Citybank, New York Life, Blue Cross/Blue Shield as well as government offices are discovering the tremendous benefits for both company and employee. It is estimated that by late 1998 60,000 federal employees will be telecommuting saving taxpayers millions and reducing traffic and smog. This in turn will benefit small towns and communities with the de-concentration of the labor force.

Advances in technology by telephone, fax machines, modems, Internet, satellite and video conferencing permits workers to perform their job duties and tasks from virtually anywhere. Tasks such as collections, billing, telemarketing, inventory control, and answering 800 calls can all be performed from home with some workers just checking in once a week a nearby "telecommuting centers".

"Work will move to unconventional sites and arrangements. Employers are becoming more willing to consider almost any work arrangement that will get the work done at less cost. Businesses are seeking to contain costs and responding to the need for flexibility". The Futurist Magazine.

Employers not able to pay more because of cost control concerns can keep their existing employees and offer new employees this added benefit of flexibility instead of a higher wage.

The popularity of this new job "perk" seems to coincide with the way Americans are feeling today regarding their jobs or careers. A recent Gallup Poll showed that 43% of women 26-45 will reduce their commitment to their job over the next 5 years with 23% excepted to quit altogether. While 33% of men will reduce the time they spend at work.

These new working arrangements seem to be a win - win situation for both employer and employee. Employees spend more time with their families, save on work related expenses such as clothing and food and also benefit their nerves and the environment by staying off the freeway and cutting down on smoggy emissions. Companies are benefiting with a

20% - 50% increase in productivity, 0 absenteeism, low turnover, savings in rent, utilities, insurance recruitment and training.

The top telecommuting jobs for the 1990's are:

- Accountants
- Lawyers
- Stockbrokers
- Marketing and Public Relations Consultants
- Business management consultants
- Environmental consultants
- Political Consultants

Big Business In America's Backyard

The availability of good jobs in America's small town's and rural areas is looking up. American companies, released from geographic restraints by new technologies and a completed transportation system are increasingly opting for rural locations. These areas offer more amenities for employees along with lower costs.

"Increasingly, in order to market products and services, companies have to look at niche markets. That tends to lower the size of some economic units, among them autonomous units that can be located in rural areas. Small companies and small units of larger firms tend to favor the advantages offered by rural communities". Stated Russel Youmans from Oregon State University in the September/October issue of Expansion Management Magazine.

The advantages a rural community can offer a company looking to relocate includes:

- Lower Wages. Because of lower cost of living expenses, rural workers receive up to 25% less.
- Cost of real estate and space rental is cheaper.
- Lower Utility Rates.
- Lower Tax Rates.
- Room to expand. Buildings and space to park.
- A better work ethic. Rural workers want to work and are more productive.
- Improved amenities for the employer and the employee. A more positive environment. More affordable housing, lower cost of living, safer neighborhoods, better schools and less pollution.
- New rural businesses have a higher survival rate than those of new businesses in metro areas.

Some companies find it just to expensive to stay in major metropolitan areas today. New York City exemplifies the demise in the number of urban-based businesses.

PERCENTAGE OF FORTUNE 500 COMPANIES IN NEW YORK CITY

The causes for business leaving metro areas simply stems back to the reason, these areas are old. Deteriorating buildings. Most companies find it to difficult and cost prohibitive to update, retrofit or expand old buildings. These older business districts also carry with them an overabundance of other problems. The majority of problems are social. Inner cities have a poor labor pool and concerns over crime keep employees feeling unsafe in the work place.

Lower land costs are the main attraction for many companies looking to relocate to rural areas. Particularly companies requiring a large amount of space such as distribution centers. A leading retailer, Best Buy, recently located one of its regional distribution centers to Findlay, Ohio, population 30,000, due in part to it's low land costs. With one of the worlds most complete transportation systems and low land costs, large distribution companies can locate in just about any rural area and thrive.

Most existing companies in rural America are manufacturing. About 700 of the over 2000 counties are based in Manufacturing. Followed by Service, Retirement/Tourist and Farming.

Long Standing Rural Employment

Much of rural employment has remained constant in certain areas. Although the rural rebound movement has created more job opportunities there has always been the good old stand by's in small town employment:

State and Federal Jobs: Many government agencies have offices located in small towns. The United States Department Of Agriculture employs thousands of people in rural areas. Their offices are set up to assist farmers and ranchers and are generally located in small towns to better serve the agricultural communities. State and Federal prisons are usually always located in rural areas for security purposes. A large portion of rural America still lives in poverty and requires services such as welfare and food stamps. These programs are administered by a state and federal staff located in rural communities. Truck weigh stations and border inspections tend to be located at rural sites. Rural offices for the Department Of Motor Vehicles and Drives Licenses also employee local residents.

County Level Jobs: Rural counties, like big metro cities, provide services to their residents. Jobs can be found at the local Sheriff's Office and Jail, County Recorder, County Nurse, school district, Building Department and Assessors Office.

Hospitals and Nursing Homes: In many rural areas the largest employer can be the local hospital. They usually tend to pay good wages since they are billing insurance companies and well, you know the story there. Many nursing homes are located in rural areas due to the fact it is just simply cheaper to maintain these homes on rural operating costs. With many older Americans only relying on social security benefits, many of the homes in rural areas have flourished, offering the patients more affordable rates than being placed in the city.

Retail: Retail establishments such as grocery stores, restaurants, gas stations are usually the main source of employment for small towns. But they tend to pay employees very low wages and no benefits.

	Farming	**Manufacturing**	**Service**	**Retirement**
Earnings Per Job 1989	**$18,570**	**$18,950**	**$18,023**	**$17,603**
Median Family Income 1989	**$24,394**	**$26,936**	**$27,677**	**$26,657**

Rural Home- Based Business

Many individuals looking to start their own home based business are frequently turning to small town America and rural areas as a prime location in which to operate from. The main motivation for these individuals is financial. A lower cost of doing business. In rural areas you find that the labor is cheaper, lower rents, lower office and warehouse rentals and a lower cost of living for the owner. Setting up an office at home or in a rural area can save a business $5,000 to $20,000 a year in operating expenses.

30% of all home workers or 10 million Americans are self-employed with the top ten home based businesses being:

1. Consulting
2. Computer Services/Programming
3. Business Support/Services
4. Financial Support/Services
5. Independent Sales
6. Graphic, Visual or Fine Arts
7. Writing
8. Marketing/Advertising
9. Construction/Repair
10. Real Estate

Source: 1991 Home Office Computing Magazine

Today, all top ten businesses are being successfully conducted from the country. In a recent Inc. Magazine article, David Birch, President of Cognetics Inc. stated: " With modern telecommunications facilities, far reaching parcel delivery service, interstate highway system, and a US Post Office, the possibilities for growing a successful company in the boondocks are quite real." And in What's Behind Small Business Success: Lessons From Rural Communities the author asserts: "almost any type of enterprise can be developed in almost any rural American community".

Turning Your Job Into A Business

CURRENT JOB	BUSINESS OWNER
ADMINISTRATIVE ASSISTANT	DESKTOP PUBLISHER WORD PROCESSOR
ACCOUNTANT	TAX PREPARER/ADVISOR BOOKKEEPER, COLLECTOR
COMPUTER PROGRAMER	CONSULTING SOFTWARE DEVELOPER WEB SITE DESIGN
NURSE	MEDICAL TRANSCIPTIONIST CLAIMS PROCESSOR
TEACHER	HOME SCHOOL ADVISOR PRIVATE TUTOR
MECHANIC	MOBILE MECHANICAL SERVICE
CHEF	CATERER

Operating A Rural Business

Although many people successfully operate their businesses from the country not every rural or small town location is right for every business. If your planning on running your business from your home in the country concerns such as lot size, distance and out buildings may have to be taken into consideration.

Whether you plan to immediately go into business or it is a possibility in the future the following questions can save time and expense when hunting for your country home and rural business site.

1. Will your home business require customers to come to you or will you take phone/mail orders and ship via parcel post? If you business requires customers coming to you, such as auto service or a florist, distance may become a problem for your potential customers. If you build custom furniture in your barn and ship directly these issue will not be of concern to you as long as your favorite shipping company can find you.
2. The size and condition of existing or planned outbuildings should be taken into consideration. Are existing buildings adequate for the manufacturing of the widgets you plan to produce? Are the conditions of the buildings suited for immediate startup? How much money will be needed to run adequate utilities. A welding shop for instance may require higher voltage electrical service. Getting the electric company to run higher voltage to a remote location can be a very expensive proposition. A home closer to town maybe more expensive but the power maybe close by.
3. Will existing or potential employees travel the distance required to get to their employment out of your rural home?
4. Are the roads to your property well maintained. Just because you can get to your property easily in the summer maybe a different story for customers or a delivery service in the winter.
5. Do zoning laws prohibit certain activities on your property? Noise, hazardous waste and handling of materials, parking and building codes may all interfere with potential business.
6. The distance from emergency help maybe a concern if the business requires hazardous work. If an accident where to occur in a business that requires the handling of hazardous materials can you get immediate attention to employees?
7. Is property in y[illegible] county taxed different? If you run an operation out of your home or outbuildings will those buildings then be considered non-residential and than taxed accordingly?

200 Ways To Earn A Living In The Country

1. Abstracting Service
2. Accountant
3. Adoption Coordinator
4. Advertising Sales
5. Airfield Ownership
6. Animal Broker
7. Answering Service
8. Antique Dealer
9. Appliance Repair
10. Appraiser
11. Arbitrator

12. Architect
13. Art Restoration
14. Attorney
15. Auctioneer
16. Auto Repair/Maintenance
17. Automotive Loan Broker
18. Bakery
19. Bank Teller
20. Bankruptcy Service
21. Bartender
22. Barter Service
23. Beauty Operator
24. Bed & Breakfast
25. Blade Sharpening Service
26. Bookkeeping
27. Bridal Coordinator
28. Business Plan Writer
29. Butcher
30. CAD Designer
31. Cake Decorator
32. Calendar Service
33. Campground
34. Candle Maker
35. Canoe Livery
36. Car Sales
37. Carpet Cleaning
38. Carpet Sales and Installation
39. Catalog Owner
40. Caterer
41. Chimney Sweep
42. Collections Service
43. Computer Consultant
44. Computer Repair
45. Consultant
46. Cook
47. Corporate Event Planner
48. Counselor
49. County Commissioner
50. CPR/First Aid Instructor
51. Craft Lessons
52. Dance Instructor
53. Day Camp
54. Day Care
55. Dentist
56. Deputy Sheriff
57. Desktop Publisher
58. Destination Restaurant
59. Doctor
60. Dog Trainer
61. Draftsman
62. Electrician
63. Emergency Response Coordinator
64. Entertainer

65. Executive Search
66. Fan Club Manager
67. Farm Equipment Appraiser
68. Farm/Ranch Sitting
69. Fast Food Manager
70. Feed Store Owner
71. Fencing Service
72. Financial Planner
73. Firewood Service
74. Flea Market Organizer
75. Florist
76. Food Cooperative
77. Foster Care
78. Freelance Writer/Editor
79. Fund Raising
80. Furniture Builder
81. Furniture Refinisher
82. General Contractor
83. Gift Basket Service
84. Gift /Craft Store Owner
85. Glass Repair
86. Grainery Manger
87. Grant/Proposal Writing
88. Greeting Card Sender
89. Handyman Service
90. Hauling Service
91. Home Inspection
92. Home School Consultant
93. Horse Shoer
94. Horse Trainer
95. Horse/Stock Trailer Service
96. Hot Air Balloon Rides
97. Importing/Exporting
98. In Home Nursing Care
99. Incorporation Service
100. Industrial Equipment Broker
101. Insurance Agent
102. Internet Sales/Marketing
103. Inventor
104. Investment Broker
105. Irrigation Services
106. Job Hot Line
107. Landscape Designer
108. Landscape Service
109. Laundry/Ironing Service
110. Legal Typing
111. Liberian
112. Liquidator
113. Locksmith
114. Log Home Distributor
115. Long Distance Phone Service
116. Maid Service
117. Mail Service

118. Manufacturer's Rep
119. Map Publisher
120. Massage Therapist
121. Medical Claims Processing
122. Medical Transcriptionist
123. Mobile Disc Jockey
124. Mortgage Loan Broker
125. Moving Company
126. Newspaper Reporter
127. Nine Hundred #
128. Nurse
129. On-Line Job Service
130. Painter
131. Paralegal
132. Party/Equipment Rentals
133. Pest Control Service
134. Pet Breeder
135. Pet Grooming
136. Pharmacist
137. Photographer
138. Plumber
139. Pollster
140. Pottery
141. Pre Fab Home Sales
142. Private Detective
143. Real Estate Agent
144. Real Estate Appraiser
145. Recycling Service
146. Relocation Consultant
147. Resale Shop
148. Residence For the Elderly
149. Restoration Service
150. Resume Service
151. Retail Store Manager
152. Retirement Planner
153. Roofing Business
154. Salvage Service
155. Satellite Equipment Sales/Service
156. Scanning Service
157. Seamstress
158. Secretarial Service
159. Septic Service
160. Sightseeing Excursions
161. Sign Maker
162. Silk Screening
163. Small Engine Repair
164. Snow Removal
165. Specialty Food Production
166. Speechwriter
167. Store Clerk
168. Substitute Teacher
169. Surveyor
170. Tanning Salon

171. Tax Preparation
172. Taxidermist
173. Teacher
174. Telemarketing
175. Television/VCR Repair
176. Tire Repair
177. Towing Service
178. Trail Ride Service
179. Translation
180. Trapshooting/Hunting Range
181. Trash Collection
182. Tree Service
183. Trophy Engrave Service
184. Truck Driver
185. Trucking Broker
186. Tutor
187. Upholsterer
188. Vacation Rental Broker
189. Veterinary
190. Video Transfer
191. Waitress
192. Water Pumping Service
193. Weaver
194. Welder
195. Wholesale Travel Agent
196. Windshield Repair
197. Wine/Beer Making
198. Woodcarving
199. Word Processing
200. WWW Page Creator

Careers In Agriculture

Today in U.S. Agriculture there is a shortage of qualified applicants to fill career openings. The USDA recently commissioned a report that polled a number of experts from several of the top Agricultural Universities. The poll showed that from 1995 to the year 2000 there will be close to 48,000 agriculture related job openings each year, however only 45, 675 graduates will be available to fill those positions. This equates to 5% of agricultural related jobs will go unfilled.

The report showed that that greatest need would be in agriculture, environmental studies, veterinary medicine, and non-medical areas of life science research. The report notes that the best career opportunities in these areas:

- Social Services Professionals - Food Inspectors, Nutritionists, Dietitians and Recreation Specialists.
- Sales, Marketing and Merchandising
- Scientists and Engineers - Food Quality and Safety, Research and Development of New Agricultural Products.

One of the hottest job opportunities in agricultural - rural based employment will be careers in value added processing. According to the staff at the University of Minnesota-Crookston it is the "region's most exciting growth industry".

For those rural inhabitants will either agricultural degrees or experience many career options exist:

- Crop Consultant
- Range Management
- Soil Consultant
- Nutritional Consultant
- Breeding Selection/Scheduling Consultant
- Disease Management
- Feed Trail Research
- Farm and Ranch Management
- Weather Consultant
- Livestock and Equipment Appraiser

In the September 1997 issue of the Fence Post, a weekly Ranching and Farming Periodical, a agricultural job service claiming to be the nations largest, AgJobs USA advertised the following agricultural related salary ranges: $20,000-$50,000 for beef, farming, dairy, pork and agribusiness. Many $25,000 to $35,000 available. Website address: www.agjobsusa.com

Another placement service, Hansen Agri Placement based in Grand Island , Nebraska, advertised the following:

Crop Specialist	$30,000
Working Farm Manager	$30,000
Ranch Manager	$30,000
Custom Applicator	$33,000
Ranch Hand	$18,000
Ranch Foreman	$30,000
Equipment Operator	$30,000

Making A Job Out Of Living On Less

Even though Americans are among the wealthiest people in the world, with a median annual income of more than $31,000 per year, more than 1 million Americans file for bankruptcy every year. We owe $1.195 trillion. We possess more than 1 million credit cards. The average family owning 3-4 major credit cards plus 5-6 store cards. 19% of our disposable income goes to paying credit other than our mortgage.

The average per-household consumer debt:

1990	1995	2000
$38,734	**$50,529**	**$65,726**

On top of those sobering statistics more of us are working longer and making less. 60% of the American Male Workforce saw their wages fall since the 70's. Median Income has also fallen, despite 2 paycheck families. With debt at an all time high and wages falling many workers find them selves working longer hours just to make ends meet. The old saying, "Living to work not working to live" is their reality.

A recent CBN news story cited government studies that showed women working outside the home are costing the family about $70.00 a month. Child care, wardrobe, transportation, meal expenses and placing the couple in a higher income tax bracket eats up all her income plus places the family in the red an average of $70.00 a month.

"For many people lack of time is the new poverty". Says Robert Heilbroner, professor of economic of the New School For Social Research in New York City. This epidemic of overwhelmed, underpaid, indebted lifestyles has created a sort of backlash against the popular yuppie lifestyle of the 1980's.

Implementing frugality and a cost of living adjustment seems to be the only way many Americans can crawl out of this bottom less pit of debt. The Voluntary Simplicity Movement, books like Is Less More? and newsletters such as The Tightwad Gazette are gaining tremendous popularity as Americans seeks to change the sense of what is valued.

In a recent survey, 72% of Americans in their 40's stated their desire to pare down their lives with 28% stating that in the last 5 years they had already made changes that meant less income.

The American Media tells us if we don't buy things, expensive things, that there is something wrong with us. It promotes impulse buying to satisfy some type of status and worth need.

Voluntary Simplicity teaches buying things you need instead of want. Reducing spending, consumption and debt and increasing time to do the things you really want to do not the things you have to do. Less time working and shopping means more time and money to spend on the things that we really care about.

Finding peace with not having a big house, not having a new car, not taking expensive trips, not buying expensive designer clothing in exchange for having more time with family and living a simpler lifestyle. The movement also argues by consuming less we preserve not only our resources, but the earth's limited resources.

Chapter 3 Buying Rural Property

Rural living is a lifestyle decision to which the benefits will far outlast any drawbacks. Having said that I would also add...rural living is nothing like living in the suburbs or the city. The transition for some can be a difficult one. Fantasies about living a peaceful, independent, existence in the country, far from the maddening crowds, can soon turn to feelings of isolation and hardship. Many people buy rural property on a weekend binge or because it's the trendy thing to do. Deliberation and exploration should be made before packing up the wagon and heading out to the prairie.

County governments do not provide the same level of service such as those in the city. Those who move to get away from urban problems soon discover they no longer have the same conveniences that living in the city provides. The transition can be educational to say the least.

One can become very disillusioned with the hard work that comes with rural property ownership. Sometimes getting to work in the morning will mean rising before the sun to clear out a 500 foot drive way after a major snow storm. And in some cases not being able to get to work at all because the small county you live in only has 2 snowplows, which are not scheduled to dig you out for 3 days.

Keeping up fence to keep your animals in or the neighbors animals out can be a monthly chore. Long trips into the nearest metropolitan area to get your favorite blend of specialty coffee or exotic dish can become cumbersome.

Then there are the varmints that may take exception to you co-inhabiting their space. Its nature trying to reclaim itself. Expect many more and different species of bugs. Spider, crickets, bees and mosquitoes at certain times of the year can become a real problem.

The antics of skunks, raccoons, coyotes, mountain lions and bears, now more familiar to people due to the effects of rural sprawl, can be destructive, frightening and troublesome. Even more so with most states enacting new laws to prohibit poisoning, trapping or destroying these animals.

When purchasing a small farm, you not only have the up keep on the house you may have outbuildings to maintain, especially if you have animals. They are very hard on both the exterior and interior of barns and sheds. Expect more painting, roofing and replacement of siding. In addition to all the latches and hinges on stall and outside access doors that get broken when a 1200 pound animal like a horse or cow decide they want out. Or when your pigs decide they are hungry or bored and the stall door looks like a chew toy to them.

There is so much to consider when moving to and living in a rural setting. The decision to buy country property can be complex. The task of buying rural property can be much more complicated than buying city property. Yet many people think, what could go wrong with just buying a piece of land? In the city you don't have to worry about water rights, boundary lines, property access and utilities. These are just a given when buying metro housing.

Many people choose to move to the country for a better sense of freedom. This freedom can be a double-edged sword. Freedom to do with your land what you want with out with restraints of homeowner associations and stringent building codes and regulations and nosy neighbors. But with that freedom also comes the added responsibility of not relying on government agencies to always provide your electricity, gas, and water or road maintenance.

Services can become a major expense in the country. If you think your getting a great deal on an inexpensive piece of rural property this maybe an indication of problems with services and access. Your great deal may turn into a great deal of expense. Wiping out any cost savings incurred on the purchase of the property.

Careful consideration should be taken when purchasing land that does not have a well, phone line, electricity or a county built and maintained road. Any one of these necessities alone can cost you in the thousands if not tens of thousands of dollars.

On the up side many of the inconveniences of past rural living have been overcome. Independent home energy systems have become more affordable. Advancement in reliability and the popularity in alternative energy sources such as solar, steam engines and generators has placed these technologies in reach of the country residents budget. The tremendous growth in the number of cell phones has brought the price down almost as low as the model that plugs into the wall. Satellite and direct TV dishes make TV reception available even in the most rugged remote mountain ranges.

Comfortable, multi purpose, 4 wheel drive vehicles and a completed highway system have made many rural areas more accessible. Mail Order Companies can deliver food, clothing, auto parts, books and just about anything required for daily living right to your door via one of the many highly competitive shipping companies. Progress has made country life a real possibility even for the staunchest urbanite.

Location, Location, Location

The first step to realizing the basic direction and vision for your rural property is determining in what area of the country you would like or are required to live. I say that because this will depend on the individual circumstances. If you have an existing job in the city and have worked a deal with your boss to telecommute, coming into the office only once a week, you may only be able to look at country property within 200-300 miles of your place of employment.

A retired couple maybe able to pick and move from any location in the nation. Very independent, adventurous individuals who have saved up a grub stake from large equities, severance pays or company shares may decide like those at the turn of the century to head out across the Plains in search of a potential homestead. Medical and family concerns may require a location closer to their existing location or a metropolitan area. Would be entrepreneurs may also limited by the logistics of the type of business they want to conduct.

Preferences for areas in the country maybe affected by the weather and topography. Amounts of extreme cold and heat may affect certain medical conditions such as Asthma, along with altitude and the amount of moisture in the air.

Interests and hobbies play can play a major role in where a person chooses to live. If you like to ski you would prefer a mountainous state. If one were an avid gardener they would prefer a state with a longer growing season and not a lot of restrictions on irrigating. Deep sea fishing and surfing would require living in a coastal state.

Whatever the reason the United States contains just about any type of climate and topography to choose from.

State	Climate / Topography	State	Climate / Topography
ALABAMA	HOT – HUMID PLAINS-MOUNTAINS	NEW HAMPSHIRE	COLD COSTAL-MOUNTAINS
ARIZONA	HOT – DRY PLAINS-MOUNTAINS	NEW JERSEY	FAIR COSTAL- MOUNTAINS
ARKANSAS	HOT – HUMID PLAINS-MOUNTAINS	NEW MEXICO	HOT – DRY PLAINS-MOUNTAINS
CALIFORNIA	HOT - DRY – FAIR COSTAL-MOUNTAINS	NEW YORK	FAIR COSTAL- MOUNTAINS
COLORADO	FAIR PLAINS-MOUNTAINS	NORTH CAROLINA	HOT – HUMID COSTAL-MOUNTAINS
CONNECTICUT	FAIR COSTAL	NORTH DAKOTA	COLD PLAINS
DELEWARE	FAIR COSTAL	OHIO	FAIR PLAINS
FLORIDA	HOT – HUMID PLAINS-COSTAL	OKLAHOMA	FAIR PLAINS
GEORGIA	HOT – HUMID COSTAL-MOUNTAINS	OREGON	FAIR COSTAL-MOUNTIANS
IDAHO	FAIR PLAINS-MOUNTAINS	PENNSYLVANIA	FAIR PLAINS-MOUNTAINS
ILLINOIS	FAIR PLAINS	RHODE ISLAND	FAIR COSTAL-MOUNTIANS
IOWA	FAIR PLAINS	SOUTH CAROLINA	HOT – HUMID COSTAL-MOUNTIANS
KANSAS	FAIR PLAINS	SOUTH DAKOTA	COLD PLAINS
KENTUCKY	FAIR PLAINS-MOUNTAINS	TENNESSEE	HOT – HUMID PLAINS-MOUNTAINS
LOUISIANA	HOT – HUMID COSTAL-PLAINS	TEXAS	HOT – DRY PLAINS
MAINE	COLD COSTAL-MOUNTIANS	UTAH	FAIR PLAINS-MOUNTAINS
MARYLAND	FAIR COSTAL-MOUNTIANS	VERMONT	FAIR PLAINS-MOUNTAINS
MASSACHUSETTS	FAIR COSTAL-MOUNTIANS	VIRGINIA	FAIR COSTAL-MOUNTAINS
MICHIGAN	COLD PLAINS	WASHINGTON	FAIR COSTAL-MOUNTIANS
MINNESOTA	COLD PLAINS	WEST VIRGINIA	FAIR PLAINS-MOUNTAINS
MISSISSIPPI	HOT – HUMID COSTAL-PLAINS	WISCONSIN	COLD PLAINS-MOUNTAINS.
MISSOURI	FAIR PLAINS-MOUNTAINS	WYOMING	COLD PLAINS
MONTANA	COLD PLAINS-MOUNTAINS		
NEBRASKA	FAIR PLAINS		
NEVADA	HOT – DRY PLAINS		

Country Property Values And Pricing

No matter how many sunny days an area boasts or how wonderful it would be to live in the Mountains, financial limitations always play a major role in the purchase of real estate.

Increasing demand for rural property along with a diminishing supply is driving prices of rural properties up. The average value of farm property was at an all time high in 1995 at $832.00 per acre according the USDA. The increase in rural employment is attracting more people than ever to sell out and move to these locations. Even though city prices continue to remain leaps and bounds above that of country property the old supply and demand law is dictating the price to move upwards.

Demand is up not only because of the current city to rural migration but because many people are also planning a move in their future. Many individuals in their 20-40's are choosing to retire in the country and are purchasing now when the prices are low and supply is high, hoping to have the real estate paid for by retirement. They plan on living in their current urban home and making payments on their country investment until which time they can make the changeover. Many experts believe the long time trend of city home prices being more expensive than that of country home prices will reverse itself in the future._

The price of your ideal country home and the monthly payments you can afford will play a large part in the type of country place you end up with. Pre-qualifying before any attempts to look at country property would be prudent. Finding your perfect place in the country and then finding out you are unable to make the payments can be a disappointing situation. Your lender will really have a something to say in this if you want to move to your rural property with no job or will be changing career fields. However, placing 30% or more down usually results in calming the fears of your lending institution.

What type of property you end up with will be dictated by the direction in which you have decided to take your property. Generally there are four types of rural property:

Small house - Small Acreage
Large House - Small Acreage
Small House - Large Acreage
Large House - Large Acreage

If you cannot afford much, a small 2 bedroom on 5 acres where you can have a horse and some chickens maybe what is in store. An affordable mini ranch with a 4 bedroom 2 bath home with enough outbuildings and 60 acres of property to start a part time hog farming operation may be what you are after. Or fulfilling a long time dream of living and ranching at the Ponderosa meets you needs. Whatever the circumstance rural properties of all shapes, sizes and price ranges are available through out the United States.

Colorado:	64+ acres, mountain views, 8 miles from town $28,800
New Mexico:	3 acres, underground power, heavily wooded $15,000
North Dakota:	2 BR 1 BA on 24 lots with trees, fixer upper $10,000
Oklahoma:	2 wooded tracts overlooking creek $400 an acre
Oregon:	82.25 acres, springs, pasture, timber $99,000
South Dakota:	$100 per acre and up. Land with house $3775
Wyoming:	160 acres, $190 down, $259 per month, $27,990 full price

Source: Rural Property Bulletin Sampler

Location will also affect the price. The further out you go the more the price drops. If you're after large acreage you will most likely end up pretty far out unless you can afford a high per acre price. Most large land parcels close in are priced to attract Developers and Builders. The Developers in turn split the property into smaller tracts, put in improvements than sell these 5 to 40 acre tracts for more ɔr 4 times what they paid for them per acre.

Expect to pay more per acre the smaller the parcel. The larger the acreage the more the price drops. Large acreage is usually sold in sections. A section being 640 acres.

Characteristics of the property can also have its effect on the price. Steep, rough terrain and poor access can influence the sales price. Many developers are unwilling to buy these properties, which will drive the price down. Property to far out can also be considered undesirable. However if you are not planning on living on your parcel for awhile this maybe a wise investment. These properties are generally sold and financed by the owners. They offer prices way below market value, little to no down and very flexible owner carry terms. In 10 years time civilization can migrate closer to outlying parcels making them much more valuable in the future.

Certain times of the year may affect the purchase price. Acreage owners are more willing to deal in the winter. The market tends to be slower with fewer lookers. Farmland in the spring and summer or harvest time seems more valuable to the owner because it is producing income. Less valuable when it is sitting idle. More repairs and access issues, such as snow removal, tend to cost the current owner more in time and money in the winter.

Sizing Up A Property

You determined the area and the price range for your new rural home. Got with a Realtor and spent a weekend looking at several options. Bare land with the idea you may possibly build and acreage with homes and outbuildings already in place. Carefully examining and closely scrutinizing the property and buildings can save major expense and burden in the long run.

Falling in love with the aesthetics of a country property is not wise. Remember you are not buying a tract home in the suburbs. You will not have the builder to fall back on when things go wrong. Many of the older rural homes in America today where built without permits. Shortcuts made in order to save time and money with the structure, electricity, plumbing and water supply were common place before the 1970's. For instance the dishwasher and washing machines in the homes in my area are plumbed to just dump the wastewater into the plants in the backyard. That was common practice in the 50's.

However, if you're handy with a hammer and screwdriver and can bring the structures up to date you could stand to make money in the long run. If you are not the home improvement type of person you may require the help of the local handyman. They are, for the most part, very reasonably priced in rural areas.

To avoid costly repairs in a newly purchased home in the city many mortgage companies require a home inspection. Since most home inspection companies are unfamiliar with rural properties it may not be cost effective or useful to hire one. If the seller is game and it comes out of his pocket this maybe an option. It couldn't hurt. However, finding one willing to drive out to the country may take some doing. If he is a city home inspector he will probably not recognize many of the common problems associated with rural homes.

Acc ess

Many people go home shopping in perfect weather. Why go in bad weather? Risk getting stuck in the snow or mud, trample in and out of houses with muddy shoes and possibly not be able to get around the acreage to examine boundary lines. To me this would be like marrying a woman without ever seeing her in the morning without her make up.

When winter sets in and bad whether comes around things may not look the same. You may not have the same access to your place as in perfect conditions.

Poor access can cause many obstacles:

Legal problems may arise with ingress and egress. Easements to gain access to your property through another person's property may require negotiations and the filing of deeds and contracts.

Emergency response times can be affected with poor property access or hidden, unmarked properties. Large fire trucks may not be able travel down poorly maintained or narrow roads. Roads that are to steep can prove to be impassable when covered with ice.

Maintaining a private road can require large expensive machinery. Most access road surfaces are gravel, dirt or crushed rock. Heavy rain and snow produce ruts and holes that need to be graded occasionally and smoothed out. Building a new access road will need to be properly engineered and constructed and can prove to be a major expense.

School buses only drive on county maintained roads. You may have to drive your children to the bus stop and wait in the car for the bus to come along.

In bad weather private and county maintained roads may become impassable requiring a four-wheel drive vehicle with chains. When the snow melts off expect a muddy vehicle both inside and out until things dry out.

Dust from road graders regularly maintaining the county roads, roads being closed for the removal of dirt or addition of dirt to the road and the noise from the grader can be a monthly to quarterly problem depending on how traveled the dirt road you live on is.

Boundary Lines

The only way to truly know where the boundaries are to a property is to have it surveyed by a licensed surveyor. This can be expensive and usually the seller will pick up the cost. Or if the current owner were in possession of an existing survey it would be wise to require it be turned over at the real estate closing. Taking someone's word for where boundary lines are can be a costly mistake. Especially after your fence is installed. Best to avoid having your neighbor pay you a visit with his survey, after you have installed a mile of very costly fence, only to have him tell you it has been installed on his property and he would like you to promptly move it.

Disputes over boundary lines and fencing long ago spurred most states to enact fence or "open range" laws. In most states you have a fence not to keep your animals in, but to keep someone else's animals out. Most rural property owners are not aware of this. Fences are meant to keep roaming animals out not in. In other words if you do not have a fence up, or a properly constructed fence per state guidelines, and

Farmer Johns bull wanders into your property and takes out everything in his path; you have no recourse against Farmer John or the bull.

In the State Of Colorado the law goes something like this: "any person maintaining in good repair a lawful fence may recover damages for trespass and injury to grass, garden or vegetable products or other crops of such person from the owner of any livestock which break through such fences". "No person shall recover damages for such a trespass or injury unless such grass, garden or vegetable products or crops were protected by such a lawful fence." According to Colorado law a lawful fence would be: "a well constructed, three barbed wire fence with substantial posts set at a distance of approximately 20 feet apart, and sufficient to turn ordinary horses and cattle, with all gates equally as good as the fence or any other fence of like efficiency".

The same laws apply to partition fences between adjoining properties and are also subject to more fence laws. These laws require maintenance by both owners. Colorado law states "Partition fence between agricultural and grazing land shall be erected and kept in repair at the joint cost of the owners or the respective adjoining tracts."

Laws prohibiting fencing of rivers, streams and flowing water are also in place in most states. To avoid costly litigation rural property ownership also brings with it some legal research and a willingness to stay on top of current and changing property statutes.

Zoning

Most rural property is zoned Agricultural. This zoning was originally intended only for "working" ranches and farms but in most states has been extended to small acreage rural residential property. In most states there are benefits and drawbacks to being zoned agricultural. Sometimes there is a "catch" to keeping your agricultural status. A check with state and country requirements can assist with retaining ag zoning.

The benefits to Agricultural Zoning are:

Being able to construct more than one home on your property. Most states allow for bunkhouses or a caretakers home, which in most cases can be just as large as the main house.

Lower property taxes. Most state tax has a different tax rate for urban and agricultural properties.

Less rigid building regulations. Building structure limitations are not as stringent allowing for taller buildings or the placement of mobile or modular homes.

Allows for more usage's of the property. Usage that would not be permitted on an urban property such as shooting ranges or bed & breakfasts.

Drawback to Agricultural Zoning:

A minimum lot size and restrictions on subdividing.

Some states and counties require the owner to keep a small heard of livestock or crops to retain their agricultural status.

Many states have "Right To Farm" zoning laws in place and new ones are being enacted all the time. This law provides for landowners in a rural setting to accept the sights, sounds, smells and activities of farmers and ranchers. These laws provide for the fact that agricultural operations are a normal and necessary aspect of living in the country. And that non-farming and ranching property owners cannot consider them nuisances as long as the farmer and rancher operate in a responsible manner.

These laws also provide for obligatory practices on the part of rural landowners such as weed control and livestock/pet control. Many new country property owners do not realize that old laws are still on the books providing for the destruction of any animal chasing livestock. That is until the Rancher next door shoots Suburban Fido for chasing his cattle. Under the law he had the right.

If your new country property is located in a fast developing community new roads, drainage ditches, power and phone lines, schools and other state owned property development is probably in the works. Most rural homeowners do not realize that the state reserves the right of Eminent Domain. The state can take possession of any property it sees fit for its need by compensating the owner. A quick check with the local planning board will ascertain whether your property will be effected.

Mineral Rights

Some of my fondest memories of childhood were visiting my Grandfathers Farm in Pennsylvania. One of the memories includes falling asleep to the sound of rumbling coal trucks going through the farm in the night. My grandfather had given the coal company rights to cross the property and mine for coal. This was a legal and binding contract that if he chose to sell the property would have been passed to the new owner. Previous owners of property can sell or reserve the right to the minerals, water or timber on the land. The existing owner, real estate owner or county should be consulted on this matter before purchasing. It is very important to know what minerals maybe located under the property and who owns them.

Water

How could something as simply as water be so complicated? Because it is a necessity of life and is in short supply. More and more regulations and rules are being created all the time to dictate its uses. Also, because many of us partake from the same water source, the general good of all must be protected. Every living thing, plant or animal needs water to survive. The average human uses about 70 gallons of water at day for drinking, bathing and such. A horse about 10 gallons a day and a 500 square foot garden will use about 35 gallons. The lack of a good water supply can make any property absolutely worthless.

Wells

For the most part, generally all country homes are supplied by a well. The water is supplied to the house via an electric pump and storage tank. In most states a well permit is required and restrictions are placed on the use of the water. Wells used for ordinary households can provide for the watering of domestic livestock and usually less than one acre of irrigation.

In states like Colorado, that has experienced a shortage of water, well depth restrictions must be adhered to. Because of the depth required this could be a major expense to the property owner considering building. Deep wells required by state law can cost anywhere from $5,000 to $10,000 dollars.

Testing of the well for pesticide contaminates, especially if the land was farmed recently, would be advantageous to the potential owners. Nothing will bring down the value of a home faster than a contaminated water source. The state or the county health department for a very nominal fee can do testing.

The age of the house will probably indicate the age of the well. Unless there has been a new well punched. Before the 1970's most home wells were constructed with galvanized steel casing. That is the hole going down to the well is lined with steel. After time the steel rusts and the well can start collapsing. Requiring the well be re-cased with a new non-corrosive plastic liner or a new well dug. This can be a major expense. For a nominal fee a well company can pull the pump out of the well and make a thorough inspection.

Ponds And Reservoirs

It is a common misconception by rural property owners that they can do as they please with their property. This includes building ponds and small lakes on their site. Hefty fines and penalties can be assessed for not obtaining permits and adhering to strict engineering guidelines for the construction of ponds. In some states taking a front-end loader, making a large hole and exposing ground water carries with it a substantial fine.

Surface Water

Most states provide for the fact that all running water belongs to that state but is available for the benefit of it's citizens. Therefore property owners with streams and springs running through them do not have the right to stop the flow of water. Some water may be diverted into storage but the majority must be allowed to flow freely.

Water rights generally go by date. The oldest dated permit gets the first right to the water. During certain times of the year when the water level is high there is generally enough water for everyone. But as the water level drops property owners with older permits get priority over newer permits.

If you are purchasing property with existing irrigation ditches you must educate yourself on their usage. Before turning any water on or off in your ditch it is advisable to check with the other owners of the ditch to ensure the coordination of your water usage with your neighbors.

If you have a ditch running across your property there is a good possibility someone else owns the ditch and has the easement right to bring in heavy equipment and maintain it. These easements should be recorded somewhere.

Electric & Telephone

In the city utility services are a given. Most people are unaware of the cost the builder has incurred to run these improvements to subdivision. These are cost that are passed along in the price of the home.

City folk looking a rural property are often impressed with the low cost of unimproved parcels. Don't let this fool you. Unless you want to look into unconventional means of energy you will be charged per foot by the local utility company to bring in electricity and phone from the nearest lines, which in some cases can be miles away. Your cheap parcel just became rather expensive. It is also important to verify that the property easements are provided to allow lines to be built on your property.

Electric power may not be available in 2 and 3 phase. If you have special power requirements, it is important to know what level of power service can be provided to your property.

Because electrical equipment tends to be older in rural areas it is not as reliable as in the city. It is important to know that power outages occur much more frequently in the country than in the city. A loss of power will shut the electric pump down in the well leaving you without a source of water and groceries in the refrigerator or freezer may go bad. It's a possibility you will need to be prepared to be able to survive for many days without electricity in the country.

Technology in phone equipment in rural areas can be outdated. Running a second line can be very expensive. If you are in need of a separate fax line or a second business line be aware that it may not be available or may cost a bundle. Some areas may still only offer party lines. If you are in a mountainous or very remote area you may not be able to use your cell phone.

Heating

There are several options in heating your home. You may decide on one or like most country folk utilize a variety. 3 or 4 different sources can help in keeping the high cost of heating your home down.

Propane: Prices vary with supply and demand. Prices can be as low as 58 cents a gallon in the summer and as high as $1.30 in the winter. This is a drastic variation. The best bet is to buy or rent as large of a tank as possible and fill it in the summer when the price is low. A thousand-gallon tank will cost $580 to fill in the summer and $1300 to fill in the winter, a difference of $720. With an alternative source such as a wood stove an average family home of 1200 square feet should be able to make it through the winter on a full tank.

Wood: New designs and technologies in wood stoves have proven to be a very efficient way to supplement the heat in your home. I say supplement because it can be very cumbersome and time consuming to use the stove as a main source of heat. Better control of the heat output has made the stoves burn more efficiently. Wood stoves can now keep a home at an even 70-degree temperature for extended periods of time. Burning wood can provide cheap heat depending on the source of the wood. Heavily wooded lots, managed properly can provide years of a consistent source of wood.

Electric Heat: Electric baseboard heat can be a very clean, safe source of heat but very expensive. It is not unusual for a country home to have an electric bill of $400 to $500 a month in the dead of winter.

Heating Oil: Many country homes use heating oil as a back up heating source. I have seen portable heating oil burners sell for as low as $69. These portable units can heat up to 1200 square feet and are set on wheels so they can be moved from room to room. When propane prices go way up many individuals switch their heating source to burning oil.

Sewage Disposal

If you have always lived in the city you probably never gave much thought to the process of flushing the toilet. In the country you will be the proud owner of a septic system. In the city waste leaves the house and goes to a water treatment plant where it gets treated then sent off to the ocean. In the country your waste is your responsibility.

Technology in septic systems has not changed much over the years. If you have an older home the way your system works will be about the same as new ones being installed today. Waste flows into your underground tank from the house. The tank is a place for the waste flow to slow down and allow the solids to settle and the remaining fluid to flow into the leach field where it percolates into the soil, which cleanses it.

If your septic tank is not property installed and maintained it can cause waste to drain into the ground water. This can contaminate you and your neighbors water source. Sewage contamination of ground water can cause serious disease such as dysentery and typhoid. Proper maintenance is crucial to keeping the unit it top condition. Poor maintenance can cause clogging and backup.

As in well construction, most septic construction over 30 years old was done with galvanized steel and will eventually rust and collapse. Most well companies also install septic systems. They can inspect these systems also.

Trash

Unless you live near a small town you will not have trash service. The other options you have for removal of trash are storing it until which time you can take it to the county dump and pay a fee to dispose of it there. Many counties still allow you to have dumpsite on your property that usually just requires a large hole. Or burning you trash, which will require a permit from the local fire department.

Structures

The general condition of the house and outbuildings, even in new construction, should be closely scrutinized. The condition of improvements to land plays a large role in determining the worth of any piece of property. In newer homes disreputable builders can cause severe problems by trying to cut to many corners or through general incompetence. Many older homes were built without permits and were funded with cash, not a mortgage. When you are building a home with cash, efforts to adhere to a strict budget may have resulted in improper building practices.

Some conditions are just normal wear and tear. Older homes have "settled" from the original placement. The soil may have compacted under the house and that has caused some movement. Some of this is normal and is to be expected. But you don't want your dream mountain home to be slowly sliding down the hill.
If the foundation has severe cracks or shows signs of upheaval, or if the walls and ceiling have cracks or are separating, questions should be asked and the possibility of needing an inspection by a structural engineer should be raised.

In older homes you will want to check for wood rot buy taking a sharp object, such as a nail and poke them into beams and exposed wood. If it goes in easily
and feels mushy the wood is probably rotted. The type of nails and construction can give away the age of the home. Squared headed nails and blunt wood screws as opposed to round head nails and tapered screws where used before the 1900's.

Ceiling stains in rooms and closets can be an indication of a bad roof or plumbing leaks. Stains on the walls of the basement can suggest flooding has occurred. Dark coloring around outlets can indicate electrical problems.
Sheds and outbuildings on older properties are often in ramshackle condition. However, it is amazing what can be saved with some shoring up and new siding and roof.

A though inspection of the structures will give you a rough idea of the expense your in for within the first years of owning the property. You will be able to plan a repair budget that may require you to put less down on the home and put the remaining aside for restoration or replacement. At least it will give you some bargaining power when it comes to the ultimate price you will pay for the property.

The type of building material used in the construction of country homes can greatly affect what you will be paying for insurance. Cinder block or brick homes may provide for a substantially lower rate than that of stick built. If there is a fire for the most part the basic structure of the home will still be standing. Insurance companies also prefer the roof be composite, metal or clay as opposed to wood.
Updated electrical and heating systems can also pay a major role in the amount of insurance you will pay. Best to keep this in mind when searching out potential rural homes.

Land Management

Land that has been overgrazed can take years to return to its natural state. Land striped bare of vegetation can be highly eroded. Horses, Sheep and Cattle allowed to graze to long on a particular piece of land can do irrepritable damage to the ground. Proper management in the past and future can effect the worth of the property. There are many new land management techniques that maintain a proper balance no matter how small the acreage. Agricultural extension offices offer free information and advise as how to best utilize land in regards to livestock.

Weed control affects the value and use of your land along with your neighbors. Reluctance on the part of a rural homeowner to control weeds can cause an overabundance and seeds to blow over on anthers property. Control and abatement of certain weeds can be very costly. Many local governments provide for landowners to control weeds or face penalties.

A rural homeowner not matter what size acreage they may own are stewards of their land. Land ownership carries with it many responsibilities towards the environment and future use.

Building Sites

For the many individuals that desire to live in the country that dream maybe far off. Financial and personal limitations prohibit them from making the move now. For the time being the only option for them may be to buy a vacant piece of land now, before prices go up. The property can then be paid off and used as equity in a down payment later for the construction of a home.

When choosing bare land consideration should be made to establish a suitable building site on the property. The selection of a building site will depend on factors such as preference, financial considerations and environment.

The house is the focal point of the entire property. Sites for barns, sheds and outbuildings will also require adequate space and may need to be within distance of utility connections. However, they should be a secondary consideration since issues such as access, power and water may not be a necessity at every site.

Preference for seclus n and views may be a determining factor in choosing a home site. Heavily treed areas can offer privacy. Hilltops may offer a homeowner a breath taking view off the front or back porch. The preference for the type of home one chooses to build also decides the home site. If the building site does not allow for large square footage a two-story home may fit nicely on the home site. If you are not limited by space your property may accommodate a sprawling ranch.

Mother Nature may not condone your choice of homesites:

Trees are an aesthetically pleasing amenity and offer privacy, but can also involve the home in a forest fire.

Steep slopes can slide in heavy rains. Large rocks positioned above the house can also fall during a heavy downpour.

Putting a cottage near a small stream may not seem so quaint until that small stream becomes a raging river during heavy precipitation.

Putting a pencil and piece of paper to the financial considerations of the location of the home site could reveal many options. Of all the development expenses road construction will be the most expensive. If you want privacy, far from the main road, you will have to pay a premium price for it. Since electric and telephone line installation is priced by the foot, this will also dictate a larger expense.

After you have picked out an aesthetically pleasing site with some shade trees, view and enough level land to build the size home you desire the environment may decide whether your home site will be feasible.

Concerns over what lays underfoot on the site can affect the construction of the home. The soils absorption rate of rainwater and sewage, the underlying water table, frost depth, soil expansion, compaction and type are factors that will determine the conclusion to the site development process.

Since moving land is expensive a site which requires the least amount of leveling, blasting rocky conditions and tree removal would be the most advantageous to the pocket book.

SOIL	DRAINAGE	FROST EFFECT	RATE
GRAVEL	EXCELLENT	NONE	EXCELLENT
CLAY	POOR	AVERAGE	POOR
ORGANIC	POOR	AVERAGE	POOR
SAND	EXCELLENT	NONE	GOOD

Soil that expands and contracts and does not drain can devastate the foundation of a house. A gently sloping site that drains water and waste away from the house is preferable. If after a hard rain the ground feels exceptionally muddy or spongy this is a sign that there is poor drainage. Ground that does not drain and holds water can cause severe problems when the water in the ground goes through a fast freeze. Freezing ground expands and can crack foundation.
If you plan on gardening or farming on you site the soil is also a concern.

Overgrown ground cover can be an indication of good soil fertility. Ground that is acceptable for good crop production should have a layer of good topsoil at least 10-15 inches deep. Dark, loose, crumbly topsoil is a necessity for a bountiful harvest. Gullies, exposed roots on trees and bushes and soil that appears parched and light colored are all indications of the erosion of the topsoil.

The Environment

Outside forces in nature can affect salability of property and consequently affect the property value. Rural residents are often faced with a property that is either uninsurable or insured at a very high rate. Tornadoes, hail, heavy lighting causing fire or appliance damage and poor response times in emergency service have caused many insurance companies to pull out of areas such as eastern Colorado. Leaving a handful of companies to insure rural homes at a high rate.

Checking with an insurance company before making an offer on a home will give you a good idea what you will be paying and what you must budget into your house payment. The insurance companies that have stayed, such as Farm Bureau, offer polices that cater to small farm owners. They cover not only the home, but outbuildings, farm equipment and livestock. These are great policies and cover the effects of nature evident in rural property ownership such as severe weather and predator losses.

In 1996, more than 750 homes were lost to wild fires. Wild fires in rural areas are often times caused by forces such as lighting strikes. Controlling wild fires actually starts with being prepared for an occurrence such as this. There are many issues facing a rural homeowner with regards to fire. A limited water supply, response times from the local fire department and accessibility to the property are several to be concerned with.

Access to your home is essential under any weather condition. Driveways to narrow or to steep for a fire truck to get down can hamper rescue efforts. Turn around radius can also be an access problem. Local fire departments have specifications available for property access. Access Roads should be at least 10 feet wide. The wider the road the longer it will last because the wear and tear is spread over a larger surface. The inside radius of a curve should be 45 feet with a minimum turnaround area of 40 feet. Slope grades on access roads should be no more than 10% with culverts, gutters and ditches to control any run off. A second means of escape, another road leading in and out of your property may also be required. If the Fire Department feels they could get trapped on your property in anyway the will not respond.

Emergency response times are much slower in rural areas. For a person with a grave medical condition the country may not be the best place for them. Most rural Fire and Rescue Departments are still staffed by volunteers. That means that from the time of your call the firemen and rescue workers must drive to the fire department, suit up then respond to your property. Depending on how far out you are, this can take up to 30 minutes. If the medical emergency is severe you might want to impress that upon the emergency dispatcher and request a helicopter be sent immediately. To wait for the rescue team to arrive and then have them dispatch a helicopter can end up wasting valuable time. Most rural communities do not have large hospitals with extra services such as burn units or critical care. The severely injured, in a rural setting, will most likely be transported via helicopter to the nearest metro city.

The same access problems may interfere with police protection. Most rural Sheriff Offices drive the old stand by patrol car. These vehicles are not 4-wheel drive. If you require assistance from the Sheriff he can not help you in poor weather if he cannot reach your property. If you home is hidden from the street by brush and tress or is a long ways off the road you will want to make sure your driveway is clearly marked with your street address. Again, time for response to an emergency will be delayed. Most rural Sheriff Offices are understaffed. With one deputy possibly working at night. If it is a very small community you may have to wait for the Sheriff to get out of bed. In that respect you can consider yourself as still living in the Wild West. If an intruder is breaking into your home, whether it be a person or wildlife, you may need the protection of weapons which will require some training in their handling and care.

Water is an important element in fighting wild fires. Lack of a large enough supply of water can hinder fire abatement; trucks can only hold so much water. If their is no outside water access to well water the trucks may have to waste valuable time going to and from an adequate water source. Construction of an outside water outlet that the fire department can use may be an option. It also may mean the difference between getting insured at a reasonable rate or have to pay high rates for fire insurance.

Practices to reduce the risk of fires in rural areas should be a consideration for all country dwellers: Stacking firewood away from the house, constructing irrigated greenery around structures, keeping grass and weeds mowed down, routinely cleaning chimneys, thinning heavily brushed and wood areas.

Living with wildlife in a rural setting can take some getting used to. Birds, deer and other small creatures can be pleasurable to have around. They may seem cute and tame but can also be destructive and hard to control. That's where the "wild" in wildlife becomes apparent. There is a price to be paid for living so close to nature and can be equated into dollars and cents.

- Losses of livestock killed off by predators.
- Cost of wild animals consuming livestock or pet food.
- Repairing fence that has been affected by wild animals going through, over and under to obtain access.
- Losses to livestock and pets due to disease transmission.
- Repairs caused by rodents and birds to the structure and utilities caused by nesting.

Many new rural inhabitants think it is fun, entertaining and educational to feed and care for these animals. This can be very harmful to the wild animals, as well as yourself, your family and your domesticated animals. Many wild animals carry diseases. If they are using the same feed dishes and watering tanks as your domestic livestock and pets there is a good chance of diseases being transmitted. Rabid skunks, squirrels and raccoons can attack people and pets. Diseases carried in the fleas on these animals can also cause sickness and death.

There is also a trickle effect to trying to attract animals to your country property. If you feed the rabbits and keep the population up around your home expect coyotes. They will not only attack the rabbits but small livestock such as chickens and especially the young livestock. Even small calves will fall prey to these animals. If they get really hungry or brave they have even been known to attack small children.

The investment in a good predator control dog is well worth the money you will save in livestock. Placing livestock in barnyards at night as well as patrolling dogs will have a significant effect on your reducing your losses.

Attracting deer and large game can also pose a problem. Once they learn you are keeping a stack of good alfalfa hay on your property or discover where you keep the grain, they will be frequent visitors. The expense alone of feeding every 1000 pound wild elk for miles can get out of control fast. Not to mention the expense to fences torn down trying to get in or out of the property. Plus the danger of small children being trampled when the animal gets frightened. A fellow horse owner that I know that lives in the Rockies keeps a cowbell on all his horses. He has a problem with elk going through his fence to eat the hay he puts out for his horses. His horses, being a herd animal, follow the elk right back out after breakfast or lunch. Then the horses are lost after wandering away with the elk heard. The cowbell helps him locate the missing horses.

In many states there are strict laws pertaining to feeding, keeping, destroying and hunting of wildlife on your property. The local Division of Wildlife can fill you in on all the current laws.

By feeding wild animals you may be disrupting the natural migration. These animals are used to thousands of years of fending for themselves. The local wildlife division will take exception to anyone that interferes with this natural process. However if you are having a problem with these animals they will offer assistance, sometimes in the form of financial aid, to help you deal with the problem in a humane manner. Poisoning and trapping wildlife in most states now is strictly prohibited.

The Decision

The intent of this chapter was to give the reader a good understanding of what you are in for with rural living. It is not always easy. Those of us that were brought up in the city have become used to its many amenities. Rural life may not be for everyone. It is a trade off. It maybe a harde .estyle and more responsibility than an apartment uptown.

The decision to move to the country will ultimately come down to what you value and what you consider important. If full government services, immediate access to culture and shopping and being in the center of it all is important to you then your best bet is the city.

If feelings of safety, community and good health are important than you are probably a good candidate for the country life. One contemplating a move to the country may consider renting to preview the rural life before investing all their time and money into a new lifestyle. The more stout hearted may just take the plunge.

Chapter 4 - Understanding American Agriculture

"There are thousands, perhaps millions of people in urban situations who are unhappy because they belong in farming and do not know it. They have the true farmer's spirit in them - that blend of creative artistry, independence, manual skill, and love of nurturing that marks the true farmer. If some of these people had been exposed to intelligent and craftsmanlike farming, perhaps they might be living on and working their own little farms. And with these hundreds of thousands of carefully kept little garden farms dotting the landscape, all of society would profit". Gene Logsdon At Nature's Pace

22% of Americans stated they desired to live on a farm, according to a recent Gallup Poll. If we applied this statistic to America's working population it would equate to an interest in farming on the part of 30 million Americans.

To be sure if they desired to live on a farm, they wished to farm. The two are inseparable. You cannot own a "farm" without working it. According to Webster's Dictionary a farm is a piece of land (with house, barns, etc.) on which crops or animals are raised. Farming is a business, but more so a lifestyle.

The farm is endeared to us. The late John Denver, whose popularity was based in songs regarding the farm and country life sang, "Country roads take me home to the place I belong" and "Thank God I'm a Country Boy". Weekly television shows like The Waltons, Little House On The Prairie, Green Acres, Dr. Quinn Medicine Woman found an enormous viewing audience by depicting country and farm values. These shows are still enjoying tremendous popularity through cable re-runs. Finding a new audience among the young. Americans are still interested in the farm

Where does this lasting desire to farm come from? The desire to farm may have its origins in different reasoning depending on one's past experiences and nature. One's family history may play a large part in the urge to farm. Up until the late 1950's most of the citizens of the United States were farmers. Farming is America's heritage. To be sure American Farming and Ranching is idealized by the greater part of the world. American art and literature romanticizes the Farm, centering only on the aesthetics, departing from the fact that farming takes hardwork, guts and determination.

Many people may desire to farm because the feel an overwhelming need to be connected to the land. This fact is visible in the huge numbers of backyard gardeners in this country. Every year millions of dollars are spent on gardening magazines, books,

TV shows, gadgets, seeds and fertilizers. To fulfill the need to cultivate and care for their own special place.

A farmer has a need to be on the land and a rancher has a need to be with his animals. The Kentucky poet, philosopher and farmer, Wendell Berry writes that most people see the land as scenery, as an object, separate from themselves. On the other hand, Farmers are unable to see the farm, simply for its beauty, because they are so connected to their land. He writes:

Sowing the seed, my hand is one with the earth.
Hoeing the crop, my hands are one with the rain.
Having cared for the plants, my mind is one with the earth.
Hungry and trusting, my mind is one with the earth.
Eating the fruit, my body is one with the earth.

Ranchers draw constant criticism from animal rights activist that their practices are to be considered cruel treatment. These activists evidently have not hung around a rancher long enough to see the endless, sleepless nights staying up for births and sickness. In addition animals have been domesticated for so long it would be cruel to release them back to the wild. They would have no sense of self-protection.

A farmer's animals are a business. In order to produce good breading stock the animals must be happy, healthy and content. Baxter Black noted the rancher's view in the following..."I do solemnly swear, as shepherd of the flock, to accept the responsibility of the animals put in my care. To tend to their basic needs of food and shelter. To minister to their aliments. To put their well-being before my own, if need be. And, to relieve their pain and suffering up to and including the final bullet. I swear to treat them with respect. To always remember that we have made them dependent on us and therefore have put their lives in our hands."

Farming and Ranching can be a source of tremendous joy and satisfaction. One may never become a millionaire farming but it can bring a sense of place and independence not realized in other work. Farming is noble and idealistic. Not only is the farmer responsible for feeding the country, but a good portion of the world. Additionally, he or she is responsible for the nurturing of animals and has been given the care of the environment. Everything a farmer or rancher does must be in the interest of his livestock and land. A farmer is the guardian of his investments.

Today many people may desire to farm to escape from the complications of the modern world. People love to idealize farm life as wholesome and simple. Crime, pollution, debt and corporate greed in a disposable society have left many empty handed, that believed in the Suburban American dream. The old 1950's Television commercials that took us away from the farm promising a "better life" filled with new appliances, a swimming pool in every backyard, a car in every garage, and a lifelong, good paying job in American Manufacturing. This is no longer reality for most.

In tandem with the perceived simplicity and wholesomeness of farming comes the sense that farming represents a sense of tremendous independence and freedom. After all farmers were the first in the ranks of the self employed. Back before our current welfare system and government entanglement you took a barren piece of land and made something out of it. Nothing to fall back on just you and a belief in your abilities. And from this land you derived your livelihood, supported you family and built it into a lasting heritage to be passed to your children.

Your farm was a direct reflection of yourself. You took pride in it, you cared for it, and it cared for you. It was lasting and was later a testament to your existence and the effort you put forth in this world. Mr.O'Hare (a plantation owner, i.e. farmer) to his daughter Scarlet..."You mean to tell me Katie Scarlet that land, that Tara doesn't mean anything? Why land is the only thing worth fighting for, worth dying for, because it's the only thing that lasts".

Whatever the reason for the desire to farm the decision to farm is a complex one and must be considered in great detail. What property to buy and what crops or livestock to grow is certainly fundamental. But farming has foundation in socioeconomic decisions to. The farming arena is very political. Outside forces will shape the American small and large farms now and in the future.

To live on a farm you will be a part of the farming community. Financial, ethical and moral considerations are woven into what makes up farm fabric. It is almost impossible to get involved with farming without getting involved in farm politics. Farming is very social. Farmers are an autonomous sort that speaks their mind. Because of the hard work and risk involved, farming and ranching denotes a sense of camaraderie instead of competitiveness. Farmers want to see their fellow small farm competitor succeed.

Structured politics of government regulations and programs will become a part of the farmer's daily life. To understand what you will be in for you must understand today's farming statistics and trends. And as in every profession, when it comes to the government you are a statistic and farming is no different. A "Farm" according to the US government is "Any place which $1,000 or more worth of agricultural products are sold or normally would be sold in a year. A good Understanding of what is going on in American Agriculture and why farm programs are in place will enable a beginning farmer to qualify and benefit from them.

According to the USDA the number of small farms is expected to increase 17% over the next 10 years. The proliferation of rural residents and new financial incentives by state and federal government, has spurred a renewed interest in farming and ranching. While full time commercial farmers continue a steady decline the number of small, medium and part time farms are experiencing growth. We are returning to whence we came. America started with small farms, progressed into large industrialized farming and is now returning to small-scale farms. In large part due to the rapidly rising cost of land, equipment and inputs.

According to the 1992 census, in states like Colorado over 340,000 acres are being managed by over 18,000 different landowners, owning acreage ranging in size from 1 to 49 acres, most of whom are brand new to agriculture. If not farming as a hobby or enterprise, a large portion of the 3 million new rural residents are growing their own fruits, vegetables and livestock. Concerns over the poisons in our food, more control over what goes on our table, the desire to be self sufficient and for the most part the high cost of food in combination with stagnant wages has gotten more people back to cultivating the land.

In it's effort to re-generate interest in small farming an entire plethora of commissions, programs and incentives have been and are being formed by state and federal governments. "We cannot let America's small farm heritage, the foundation of our rural communities, just slip through our fingers. We must carefully look at the barriers facing small farms and seek solutions to these problems", stated Agriculture Secretary Dan Glickman when announcing the new national commission on small farms. "The average age of an American farmer today is 58. We need to do more to encourage the younger generation to farm, and we must continue to find ways to help small and disadvantaged producers find ways to make a decent living, keep their land, and make their small farms economically viable."

Financial incentives promoted by both federal and state governments are providing a way for a new generation of individuals to enter farming. Renewed emphasis is being placed on attracting individuals to farm after decades of the farm population falling off. What has been quite apparent, to those that have watched the crash of American Agriculture, is that the lack of reaction from the government in the antecedent years. This lack of concern on the government's part, particularly the USDA, has created a rural backlash of sorts resulting in anti-government sediment.

An array of problems has ensued in the demise of number of family farmers. It has lead to a detrimental trend of concentration in farming.

- Roughly 22 million jobs rely on farm products. 1 in 6 Americans work in Agricultural related fields. Producing more than 150 billion in renewable wealth annually. The farm connected working population is the largest group of workers in America. This group of workers, a large percentage located in rural areas, provides for the processing, transportation and distribution of food in the United States. With the decline in the number farmers came the decline in the number of these jobs. This has caused the deterioration of the nations rural communities, displacing many workers and exacerbating rural poverty.

- Large faceless agri-business companies, that have no interest in the future of farm and rural communities or the land, now dominate the farming industry. This has lead to a less competitive food market and control in supply. Consolidation in agriculture posses a threat to competitive markets. Concerns over the growing impact this could eventually have on economy has forced the federal government to take a strong look at the shake out of the family farm.

- Farm products are 15% of the gross national product with agricultural products sold in 1992 valued at $162 billion dollars. US agricultural exports hit a record 60 billion in 1996. Up 20 billion from 1990. Above computers, cars and technology Agriculture is America's number 1 export. The urban populations of the country and the world are completely dependent on farmers to bring an abundance of safe food to the table at a reasonable price. Organic food is now the at the 3 billion dollar mark. Heightened awareness of health and environmental issues, for example the ethical treatment of the land and livestock and it's effects on the well being of future generations is beginning to impact and drive this enormously important industry.

- The United States is losing over 1 million acres of productive agricultural land yearly to urbanization. In states such as Colorado, 1700 acres of farm land are being taken out of production for development each week. Total land in farms was estimated to be 946 million acres in 1992, down from 1.04 billion in 1980.

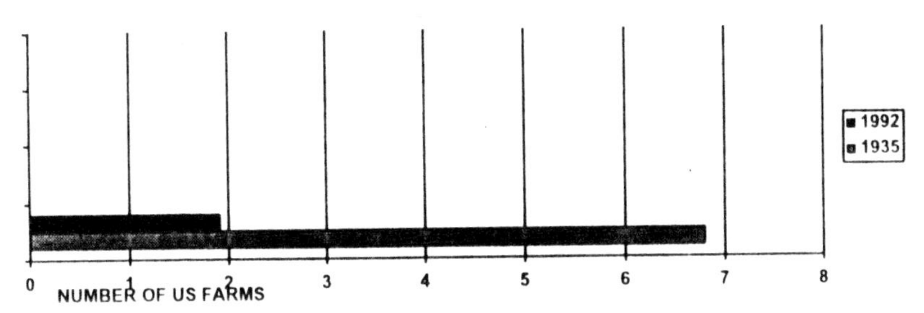

With only 2% of it's citizens remaining in farming, after decades of decline, the American farming community has reached the lowest number of farms in existence since 1850. Concern over the state of American agriculture is finally starting to get the attention it deserves. The moral, social and financial impacts rooted in the decline of farming to both rural and urban areas have made themselves apparent. A looming crisis over the family farm in which older farmers not being replaced with new ones and the impending consequences that will bring to the nations food supply have a new found importance. Financial incentives, commissions on small farms and rural affairs, public awareness and opinion seem to be favoring a change in the nation's farm structure.

There has been ongoing debate over whether farming should be a "business" and treated as such. Enduring the same economic dynamics as the rest of corporate America. Or whether farming should it be considered a "way of life" to be valued and entrusted to those recognizing the importance of safe food, care for the environment, humane treatment of animals and subsistence of their community.

The economist have put a pencil to agriculture figuring it up with the same calculations used in manufacturing and technology with no regard to social, moral and environmental implications. The proverbial chasing the tail syndrome on the issue has taken the agricultural industry in circles for years.

America needs to be concerned over who owns the farm not just what the assets of the land, machinery and inputs are. "Stressing only the business side of agriculture is inaccurate and dangerous. Many farm decisions go beyond economic costs and benefits analyses. The danger lies in the fact that the farm as a business is more easily the target of inflexible and insensitive regulations that disregard the negative impact they may have on families and communities". Sandy Rikoon Farm Journal, April 1994, Article: Wanted A New Social Contract.

The farm and food system involves more than just yielding crops or livestock. There are forward and backward economic connections. Since farmers must buy inputs such as seed, fertilizer, machinery, fuel veterinary services and transportation the farm supports the economic future of the local community. Along the way to the consumer processors, wholesales and retailers all depend on the American Farm.

In his speech to the American Farm Bureau Dan Glickman concluded that "We know in our hearts that America will not be strong unless American agriculture is strong."

The American Farmer

The 1992 Census of Agriculture states that the average American Farmer is white, 58, and lists his principal occupation as farmer. He is a full owner of his farm been there an average of 26.1 years. He will be retiring in the next 5 to 10 years with no one to take his place. Currently there are 2 farmers, over the age of 65, to every 1 farmer under the age of 35. This is in stark contrast to the rest of the countries labor force. By comparison only 3% of the civilian non-farm labor force is over the age of 65.

In 1992 the age, number and income of farmers was recorded as such:

AGE	TOTAL	INCOME
Less than 35	180,401	33,115
35-44	394,137	42,096
45-54	471,458	52,215
55-64	433,343	45,623
65-over	556,352	27,219

The Young Farmer

According to the 1992 US Census there are now fewer farmers under the age of 35 and more farmers over the age of 65 since tracking of farmers ages began in 1910.

The number of farmers under the age of 35 fell from 217,000 in 1978 to 115,000 by 1987 with the number of farms falling off by 1 million. At the height of the farm crisis in 1985, the USDA estimated that the average annual entry into farming fell 25%. "Most of the decline in entry occurred among people under 35 years of age whom made up 58% of the decline." The USDA reported in its 1993 Report Farm Finance: Number of New Farms is Declining.

Parallel to the USDA reports, every 5 years the US Commerce Department publishes a detailed Census of Agriculture. The last 2 taken, in 1987 and 1992, showed a steep drop in the numbers of young people in farming and ranching.
A recent poll taken by Iowa State University showed that today only 1/3 of Iowa's existing farmers expect their children to take over their farm.

Today's young farmer is very idealistic but also retains a good understanding of the realities they face pursuing a career in agriculture. Most young farmers plan on inheriting the farm from their families and carrying on their families history of land stewardship while others through the experience of working on anothers farm gained the longing for a farm of their own.

The are also very receptive to new ideas, particularly when it applies to being able to cut costs. Young farmers tend to farm with fewer chemicals and are generally more sensitive to conservation issues.

In it's Annual Survey Of Farmers, Farm Bureau reported a greater use of conservation and environmentally beneficial farming practices among farmers ages 18-35.

73%	Employ conservation tillage on the farm.
59%	Regularly test soil or crop tissue prior to nutrient application.
58.3%	Practice crop rotation with 3 or more crops.
1/3	Said they use integrated pest management techniques such as field scouting to reduce crop protectant use.
33%	Regularly test their well water.
22%	Enrolled land in the Conservation Reserve Program.
18.8%	Actively manage wetland resources.
18%	Leave crop strips (leftovers) unharvested for wildlife.
97.2%	Plan to be a lifelong farmer.
90.9%	Want their children to follow in footsteps.
76.1%	Plan on expanding to a larger operation within 5 years.
65.2%	Based choice on agriculture's lifestyle and business aspects.
1/3	Chose farming mainly due to lifestyle advantages.
80.8%	Are more optimistic about farming than 5 years ago
75.3%	Feel they are "better off" than 5 years ago

Farms By Race

The average age of the only 23,000 black farmers operating in the US is 58 years. Less than 1% of all African American farmers are under the age of 25. In the early 1900's African Americans owned 15 million acres of land. Today they own less than 3 million acres, the majority of which is located in the southeast. African Americans are losing their land at a rate three times higher than farmers overall. "Lack of access to federal and private credit pushes existing farmers out of business and deters young African Americans from entering into production, that's why we have only 187 African American farmers under the age of 37 in the United States". Ralph Paige, Director of the Federation of Southern Cooperatives.

Spanish owned farms faired worse at only 10,414 farms. Traditionally the Hispanic population has provided farm labor but has not represented any significant numbers of farm ownership.

Age	Total Number
under 25	134
25-34	876
35-44	1824
45-54	2030
55-64	2519
65-over	2519

At a meeting of the board of directors of the National Cotton Council Glickman declared the USDA has having "sizable civil rights problems". He told the board, "local folks, especially in the south, are using the flexibility of our farm programs not to tailor federal policies to the needs of their local community but to discriminate."

Farms by Gender

"I really believe woman can make a good life for themselves in the ag industry" Brenda Moore Of GroMoore Farm, Rush NY

The latest Census Of Agriculture revealed a dramatic shift in farm gender.

Today more than 15% of farmers and farm managers are women. Compared with 4.5% in 1970. The number of female farmers has grown 19% in the last 10 years.

A 1994 study by Wayne State University showed today's farm woman, with her abilities and talents, is worth $27,500 to the farm and puts in 68 hours a week. The same study showed that men are worth $23,700 to the farm for the same amount of work.

Women have always played an important part in American Agriculture, although few statistics are available regarding their contributions. Frontier woman where indispensable to family farms.

They performed all the tasks that sustained the basic needs of the farm. Raising small gardens and livestock and providing for all the essential necessities of the homestead. With the mechanization of farm equipment in the mid 20th century, the need for physical strength was greatly diminished and the division of farm labor in the household became more evenly divided. With gender related stereotypes disappearing women feel more comfortable and are more accepted in their ability to perform farm tasks including livestock management, fencing and operating advanced equipment, that have in the past, been solely male domain.

Most of the farms operated by women were small but are worked less hours than their male counterparts with operations of the same size. However, the female operators were more likely to specialize in less time-intensive livestock operations while male farm operators specialized in more time intensive dairy or cash grain operations.

The fact that a greater share of female operators were elderly, played a large part in the decision of what type of operation was selected. 37% were age 65 or older, which may have restricted the number of hours they devoted to farming.

Of small female operated farms their units averaged 291 acres in 1982, while farms operated by men averaged 449 acres. About 42% of all female-operated farms were smaller than 50 acres. 8 % were 500 acres or more. The average value of farm sales operated by women was $23,000 compared with $61,000 for farms operated by men.

Woman operators are prevalent among blacks and other minorities. In 1982, about 45% of all women operators were of black or other minority races in comparison with 2% for male operators. 13% of all women operators were of Hispanic origin.

Over 70% of the female farm operators were not married (probably widowed) and over 1/3 lived alone. With the aging of the farm population, farmers have died and the farms have passed to their wives. In some states half the farmland belongs to elderly widows.

Number of farm operator households headed by:

Female	Joint	Husband
111,098	432,356	1,330,318

Today the overall gender of farming equates to:

6%	Of farms were headed by women farm operators.
23%	Joint husband wife farm operators.
71%	Operated by the husband or single men.

Female Farm Operat... -
1992 census total with other jobs: 145,156

Age	Total Number
under 25	513
25-34	3824
35-44	10399
45-54	12301
55-64	15404
65-over	31003

The Landscape of American Agriculture

Once, the vast majority of the United States was in farm or ranch ground. 50% of the U.S. land area of the 48 contiguous states are currently in farms. When the country was founded 90% of Americans derived their livelihood from farming. These were small farms. In 1900, the average farm size was 147 acres compared to 491 acres today.

The vast and different topography and whether has shaped the agricultural regions of the Untied States and has provide for a variety of foods:

Pacific Region:	Wheat, Potatoes, Vegetables, Cotton, Cattle Sugarcane, Tropical Fruits & Plants, Dairy Products
Mountain :	Wheat, Hay, Potatoes, Fruits, Vegetables, Cattle, Sheep
Northern/SouthernPlains:	Wheat, Feed, Hay, Forage, Cattle
Corn Belt:	Corn, Cattle, Hogs, Dairy Products, Feed, Grains, Soybeans, Wheat
Delta States	Soybeans, Cotton, Rice, Poultry
Southeast Region:	Cattle, Poultry, Fruits, Vegetables, Nuts
Appalachian Region:	Tobacco, Peanuts, Cattle, Dairy Products
Northeastern Region:	Grains, Cattle, Dairy Products

The top four American agricultural commodities include:
World Crop Production

- Soybeans 42.7%
- Cotton 21.2%
- Corn 34.4%
- Wheat 11.6%

Spreadsheet of American Agriculture

American Agriculture is big business. It is of enormous size and significance to the economy. In 1993 Gross Farm Income was 195.5 billion dollars. Crop production alone was $84 billion.

90% of America's 2 million farms are small. Farms are categorized as small farms being 10-200 acres, family farms 200-1000 acres and large farms are over 1000 acres.

Although there is a larger number of small farms they are not as prosperous as the large corporate owned mega farms. Almost 71% of the 2 million small farms have annual sales of less than $40,000 while less than 1% of all farms have sales greater than $1 million. However many of these small farms are considered "part-time" farms and therefore draw part time income. In 1992 the average total annual sales per farm were $84,459 with the average expenses being $67,928.

Farms with over $250,000 in sales only account for less than 6% of the total number of US farms but dominate the farming output. These large farms have control over 53% of the nations crops and 62% of the nations livestock. According to the USDA's Economic Research Service, in 1995, very large farm's with sales over 1 million dollars and small farms with sales under 20,000 faired the best. Both with an increase from 1994 to 1995 in cash receipts.

The size and income of farms varies greatly across the nation. 4% of farms in the Pacific Region had over 1 million in sales. The Mountain Region had the 2nd largest concentration of large farms. Northern Plains, Corn Belt and Lake States are dominated by mid sized farms. The Appalachia, Southern Plains, Southeast and Delta regions had the largest amount of farms with $40,000 or less in sales.

Over one half of all small farms are located in the south. Texas has the largest number of small farms with over 152,000 farms reporting sales of $20,000 or under. California had over 3,000 large farms and has largest number of farms with over $1 billion in sales.

California was also the top agricultural state with 18.2 billion in farm sales in 1992, followed by Texas with 11.6 billion and Iowa with 10.3 billion.

Small Farms

The Census Bureau in 1992 found that there were over 554,000 farms of less than 50 acres. Part time farmers typically work in nearby areas or from home and tend to livestock and crops on the weekend and evenings.

"We are seeking workable ways to fit smaller-sized operations into this New World, and enable them not just to survive, but thrive in the 21st Century." Secretary Dan Glickman at small farm commission public forum, Memphis TN 7-28-97

The largest portion of farms in the United States are small. This is due in large part because the initial investment to get into a small farm is merely a fraction of a family or large-scale farm. It only makes sense you can't pay 150,000 for a combine and use it on 100 acres and expect to make a profit.

Today's successful small farm does not compete with large farms. Small farms are highly specialized and diverse and are selling products that the large farms cannot afford to sell or are unwilling to sell. They market and sell in an entirely different way than family or corporate farms for a much higher return per unit price.

In his book "How to make $100,000 from 25 acres" Booker T. Watley states "In order to make it today in farming you need to stay out of the big farm game." However, this does not mean you can't own a small scale farming operation that becomes a large specialized farm. Small farms can generate more "profit per acre" and are often far more productive, up to 6 times more, depending on the choice of products produced. Small farms are also diversified, growing usually 2 to 3 different types of outputs therefore producing year round instead of seasonally. In turn they are not subject to the outside forces such as commodities markets and the weather. In addition, small farmers almost always are employed outside the farm, full or part time. The outside income helps stabilize the farm. With today's acceptance of "alternative lifestyles" Can you not be a business woman in heels and a suit from 9 to 5 and an overall, boot wearing farmer after 6 and on the weekends?

All these dynamics add up to the fact that the small American farm has always been in the least amount of financial trouble of all farming groups. In February of 1986 The Wall Street Journal reported that in the midst of the farm crisis many New England Farms were prospering. The article stated that these producers had not taken on the heavy debt load that other farmers typically did at that time, diversified their operations and were selling their products directly to regional urban markets.

Amish prosperity also lies in low input costs, low debts and small diversified farm operations. The Amish have farmed for years without expensive machinery and chemicals. They were virtually unscathed by the farm crisis.

A large portion of small farmers started into farming and ranching in a similar manner to myself. It all starts with a garden, some chickens and maybe a horse or two. There is a tremendous amount of contentment in becoming a steward of land and animals. It seems natural to take your interest in farming from hobby farm to enterprise. Small farms are fun, rewarding and always interesting. It is a great part time income.

With daily headlines like these:

Mad Cow Disease
Alar poisoned apples
E Coli in hamburger meat
Salmonella in chicken
Red meat causes heart attacks

The organic and specialty market has grown by leaps and bounds over the past few years making it a 3 billion-dollar industry. This is where the small farm for the 21st century can step in and clean up literally (financially and environmentally). Providing safe food at a higher return. Organic foods can bring up to 50% more. People are willing to pay more for "organic" or "specialty".

Today's health conscious consumer is willing to pay more and demanding leaner, medication, hormone free meats and fruits and vegetables grown without the application of herbicides and pesticides. Small farms can provide this invaluable service to its customer.

Words like organic, cruelty free, medication free, reduced fat, free range, lean meat, fetch top dollar. Unique and novelty products such as ostrich meat, grain fed catfish and Navajo Yellow Melons also bring high profits, products that the large farms are unwilling and unable to provide.

Small farms can also derive top dollar for livestock by becoming a breeder. Most people know that top breeds of dogs with registered pedigrees can cost considerably more than a mutt from the local pound. The same is true in the livestock industry. Goats, hogs, horses, cattle and almost any type of livestock has it's elite lines with associations that track bloodlines and issue pedigrees. Pure breeds are generally considered to be genetically superior in conformation, muscle mass and overall health. Breeders can charge maximum price to other breeders or ranchers that want to improve the genetic characteristics of their existing heard or want to avoid line breeding. These animals are also sold to show at county fairs and livestock events, some to the over 1.8 million livestock 4-Her's.

Many small farms also see higher return on their production by offering what is known as end products. Cheese from goat's milk, smoked pheasant gift baskets, meat sausages, wool socks spun directly from the sheep, jelly from fresh picked organic blueberries.... are all products individuals are willing to pay more for coming "direct from the farm".

Directly selling through mail order, pick your own, retail, roadside stands, farmers markets and small farmer cooperatives all cut out the middleman and let the small farmer realize a greater profit.

Family Farms

"The fight to save family farms isn't about farmers. It's about making sure that there is a safe and healthy food supply for all of us. It's about jobs, from Main Street to Wall Street. It's about a better America." Willie Nelson

Family farms have been the corner stone of rural America. These farmers have sustained businesses and created new rural jobs in their local communities. Farmers keep small towns and rural area churches, schools, hospitals and business alive.

During the last 10 years 600 family farms a week have been lost. Since 1975 America has lost 800,000 family farms. The farm crisis of the 1970's was the beginning of the demise of family farms. Driven by the promise of big grain sales to the former Soviet Union most farmers in the 1970's borrowed to much and over extended themselves. Many farmers bought up overpriced land in the rush to immediately increase production. These grain sales never materialized. In return land prices fell up to 72%. Many farmers were washed out.

At that time banks would not refinance farmers, but turned around and sold the farm for a discounted price and lower interest rate to others mainly being big agri-business . These individuals lost farms that had been in the family for generations with the government standing by doing nothing. John Block, US Secretary of Agriculture at USDA in the Regan Administration, took note of the number of farms that were lost and declared the government should not intervene in the process. This was the same USDA that claims to be "The People's Department".

Today, 99% of US farms are still owned by individuals and family corporations/ partnerships.

Individuals	85.9%
Family Partnerships	9.7%
Family Corporations	3.4%
Other	0.6%
Non-Family Corporations	0.4%

Most of these family farms have been inherited and have been handed down through generations. And in spite of the many adversities and risks they have chosen to remain. To them farming is not just a business but it is their way of life. The average farm operator income was $ 40,223, 97% of the national average. However, this income came from various sources including off farm income and businesses. In 1993 88% of income came from off farm income. Large farmers faired better with an average total income of $153,328, with only 21 percent coming from off farm sources. Approximately 45% of operators reported farm or ranch work as their major occupation in 1993.

Today's farmer must be able to use highly sophisticated computer driven equipment and stay on top of a technologically fast moving industry......usually via the Internet. The image of the overall wearing, ignorant, "good ole boy" is a misconception. Today some farmers have capital under their control in the range of $500,000 to 1 million dollars.

Differences in the size of acreage, commodity specialization and emphasis placed on farming full time or pursuing other career options plays a major role in the financial success of the farm.

Family farms concentrate on the large-scale production of 1 to 2 crops and compete directly with big agri-business farms. Family farms have a large amount of capital expenditures. Large land parcels and high-priced equipment are used to their fullest extent. Family farms also must rely on family, not employees for harvest, repair and maintenance. Because of the high cost of farm inputs (fertilizer, seed, equipment) this farm group tends to borrow the most amount of money, there fore is usually in the most amount of financial trouble when the crop prices fall or whether turns bad. Since 1982 Prices received by farmers have only risen 7.5% will prices paid for things such as fertilizer, seed and equipment is up 23%.

Many consider family farms outdated and inefficient because the refuse to give over to technologies that may hurt their land. Family farms try to balance their concern for leaving the farm to future generations and stewardship for the environment with farm profits.

Family farms routinely place land in conservation reserve programs and practice environmentally safe farming techniques such as no till and reduced till farming, crop rotation and organic farming. After all they have to eat the food that come off their own land, drink from the well and live in the community that surrounds them. In 1994 farmers placed 36.4 million acres of their land in the CRP to reserve and protect the environment and provided the food habitat for 75% of the nations wildlife. US farmers, at their expense, maintain 170,000 miles of windbreaks, 1.3 million acres of grass waterways and use soil tillage conservation practices on 126 million acres to prevent erosion. Then they are told by the "experts" that is why they can't make a profit and don't belong in farming. When the real reasons maybe family farms can't compete against the exclusive contracts, political favoritism and wall street investors given to their giant agri-business competitors. Plus, trying to compete with the continuing growth in food imports from Mexico and Canada. Why are we importing food from places these places when we can't keep or own family farmers in business?

Although many farm groups and organizations believe that the number family farms is still in a drastic decline others believe differently. A recent Mississippi State University study concluded that family farms are alive and well. "There has been an expansion of corporate farming since the 1980's but there also has been an apparent resurgence of the family farm in some areas", stated sociologist Frank Howell of the University's Social Science Research Center.

Howell and John K. Thomas of the Texas A&M Rural Sociology Department came to this conclusion after analyzing census data from 1982, 1987 and 1990 for more than 30,000 counties.

"We wanted to track the transformation of US agriculture from the 1980's when the Midwest first began to show signs of the crisis", Howell said. "As in other business, the competitive survive." "All types of farming were affected. There has been an expansion of corporate farming, but there has been a resurgence of the family farm. In some areas, family owned enterprises are the dominant form of farming". Howell stated.

Allan Lines and economist at Ohio State states: "Farmer's balance sheets show they are on more solid financial ground and they have become more fiscally conservative after coping with the farm crisis in the early to mid 80's. Basically it's a situation where farmers have paid off their debt in the last few years, have had strong incomes, and the cash position on their balance sheets has increased."

Some farm experts would argue that with 1 in 4 farms going out of business compared to small business with a higher rate of 1 in 7, family farmers don't have much to complain about.

Corporate Farms

"Corporations are buying up land, dominating and imposing their will and when they leave there are only ghost towns and Superfund Sites".
Bill Wenzel, Clean Water Network, Madison, Wisconsin Office

In America today, corporate farming is once of the most insidious and destructive forces to afflict the nation.

The 5 major agribusiness corporations supply 1/3 of the nation's food. They control the food supply and therefore food prices. These Agribusiness Corporations have long come under attack amid accusations of careless food handling resulting in deaths, destruction of the environment, displacement of the entire rural socioeconomic culture, price fixing, monopolization of the food supply and manipulation of farm policies all for the sake of profits.

Corporate agribusiness, in the name of efficient food production, does not and will not waste valuable time or money worrying about the environmental or social impacts of their conduct.

The joint efforts of the agri big boys to so arrogantly manipulate the food structure in this nation was summed up when the son of the Chairman of the agribusiness giant Archer Daniel's Midland was supposedly caught on tape uttering: "The competitor is our friend and the customer is our enemy". From the article "Too Big or Not To Big?" Howard Chua-Eoan Time Magazine, 10/95.

Headlines like these would bring a public outcry:

Chrysler paid not to make cars.
Government picks up tab for Microsoft's overseas marketing campaign.
Wal Mart pays all workers below minimum wage...government makes up the difference.
They sound implausible but are a reality in governments politics with big agribusiness.
They are the luxuries allowed corporate farming.

Agribusiness obtains billions of dollars worth of subsidies from taxpayers each year directly in the form of payments, bonuses and indirectly in the form of cheap commodity prices.

Every year the United States allocates tax dollars, through it's agricultural budget, to export programs and income price support programs. These programs are in place to support farmers, low income families and the academic community. However the largest recipient of these subsidies under the present system are the large agribusiness corporations. These are billion dollar companies with yearly sales in the millions and are perfectly capable of sustaining themselves without taxpayer help.

Agri-business is fleecing the taxpayer under 2 types of policies; income support and export promotion. In fact 1/3 of government agricultural program subsidies go to these highly profitable corporate giants. In all told agribusiness is subsidized with 16 billion dollars of taxpayers money.

Income supports are deficiency payments and were created in the 1970's when export sales went south. At the insistence of agribusiness the Congress and the Nixon administration enacted policies that artificially lowered the price paid for crops; allowing agribusiness to purchase crops below market prices. This resulted in direct deficiency payments from the government in the form of subsidies paid out annually.

In 1994 if it cost a farmer $3.00 to produce 1 bushel of corn. On average he received only $2.20 from agribusiness at market. In order to offset the 80 cent loss the government paid the farmer 55 cents per bushel in the form of deficiency payments. Agribusiness received the product dirt-cheap and the farmer was still left with a loss.

In reality, farmers are paying retail for their inputs (seeds, fertilizer, and equipment) and are forced to sell their crops below wholesale. Since 1982 prices paid for inputs for things such as fertilizer, seed and equipment went up 23%. With the large agi-business owning most of the input companies. Plainly speaking, agribusiness has ultimate control and is making money on both ends. It gets to purchase the farmers commodities at below wholesale. But owns the input companies and charges the farmer retail for the inputs necessary to grow the farmers crops.

Corporate Agribusiness also receives government financial aid in the form export promotion. The Export Enhancement Program, or EEP for short, was created to foster new export markets for agricultural products. The USDA's Commodity Credit Corporation awards a bonus to whichever company offers their grain for export at the cheapest price. From 1985 to 1994, under EEP, taxpayers provided agribusiness with subsidies exceeding $7 billion dollars. Cargill Inc. alone has received 1.29 billion over the past 10 years. In addition thorough the "Market Promotion Program" the federal government spends millions annually to cover marketing agricultural products in the form of advertising and promotion for the top 5 agribusiness companies.

The nation's farm policy has failed. Failed to protect the very entity it was created for, the family farmer. It has lined the pockets of these large agribusiness corporations while obliterating the family farm system.

Companies, like Cargill, have not only reaped the rewards of subsidies but have been allowed to completely dominate the food industry from field to fork. The trend is called "Vertical Monopolies" and are they are supposed to be illegal in the United States. Cargill is a farmers bank, insurance company, feed and fertilizer company, grain storage company and mill, meat processing plant and transporter. If Cargill does not want to buy the farmers wheat, the farmer has no income, he then will not have money to pay the bank, which is also owned by Cargill, who in turn can take the farmer's land. Our politicians, who depend on companies like Cargill for campaign contributions, have chosen to ignore these monopolies.

Another problem that arises from vertical monopolies is the ability on the part of agribusiness to commit fraud. The recent scandal with ConAgra is a perfect example. Peavey Grain, who is owned by Con Agra is responsible for the grading and weighing of Con Agra's Grain. Is this not the proverbial fox watching the hen house?

USDA Release No. 0087.97
WASHINGTON, March 19, 1997
ConAgra PAYS $8.3 MILLION IN PENALTIES FOR FRAUD SCHEME

Agriculture secretary Dan Glickman today announced that ConAgra, Inc. one of the nation's largest food companies agreed to pay $8.3 million in penalties, after agreeing to plead guilty to federal charges of adulteration, misgrading and misweighing of grain by the company's grain division, Peavey Grain. The settlement amount included a criminal fine of approximately $4.4 million, a reimbursement of $450,000 to the USDA for unearned storage payments and expenses associated with the investigation, and about $3.45 million as compensation for criminal profits. This $8.3 million settlement is in addition to an earlier civil settlement in excess of $2 million which was reached between ConAgra and a group of Indiana farmers.

"This settlement culminates a four-year investigation by a task force headed by USDA's Office of Inspector General.", Glickman said. "The plea agreement not only serves as punishment to the offenders, it also offers the opportunity to repay farmers who were financially damaged by these criminal acts. I want to personally thank Roger Viadero and his staff, as well as those who assisted from GIPSA and FSA, for their excellent work in this case."

In addition to the charges against ConAgra, four of its former managers agreed to plead guilty to criminal charges. The others under contract and licensed by USDA to sample grain have also agreed to plead guilty to charges in this case.

ConAgra used several schemes to defraud farmers and grain buyers to increase their own grain inventories and profits. Soybeans were purposefully misgraded allowing ConAgra to pay less to the farmer, yet sell at higher rates, water was added to grain inventories, thereby adding to its weight and increasing profits when the grain was sold, and ConAgra significantly misweighted grain being sold to end users, thereby shipping less grain than for which they were paid. In addition, gratuities were give to USDA licensed grain samplers, who allowed elevator employees to substitute poor quality grain samples with better quality grain. This resulted in false grade certificates which were used to invoice end users. These acts constituted violations of the U.S. Grain Standards Act, the U.S. Warehouse Act, and the U.S. Food, Drug and Cosmetics Act.

The practice of adding water to grain was a central issue in the investigation. Because grain is measured by weight, this practice resulted in significant inventory growth. The addition of water also caused spoilage of grain in transit due to excess moisture. Adding water to the grain constitutes adulteration, which is in violation of the law prohibiting the addition of substances to food for the purpose of adding weight. According to testimony presented to Congress, water systems used for adding water to the grain were found at many of the company's domestic facilities and export elevators.

USDA Inspector General Roger Viadero said, "This is the first time we have used the adulteration statute when proving that water was added to grain. While the company initially claimed the water was added to suppress dust, our investigation proved the true intent was to make more money."

USDA's Grain Inspection, Packers and Stockyards Administration and Farm Service Agency joined OIG in this investigation.

"This investigation was instrumental in amending the U.S. Grain Standards Act to increase criminal penalties for violations." GIPSA Administrator James Baker said. "It was also instrumental in amending U.S. Grain Standards Act regulations on February 11, 1995, to prohibit the application of water to grain."
"Repayment to the government of storage payments paid to ConAgra by the Farm Service Agency also represents a historic move by the USDA," said FSA Administrator Grant Buntrock. "Storage payments are normally only reclaimed when grain warehouses fail to keep the required quantity of grain on hand, or the inventories of the required grade of grain fall short. In this instance, although this Department lost no grain, USDA and the U.S. Attorney's office for the Southern District of Indiana agreed that ConAgra had failed to live up to the terms of their agreement with USDA, which prohibited the addition of water to grain."
"This means that grain warehouses will have to fully live up to the standards set forth in their grain storage agreements with USDA," Viadero said. "While prosecution will undoubtedly be restricted to those operations which use fraud and deceit, these precedent setting prosecutions have given us a new tool with which to protect the interest of this Department, and the farmers of this nation. The American Farmer is the key to all agribusiness and the cornerstone of the American market, both domestic and around the world. USDA is dedicated to preserving the right of the farmer, ensuring that everyone receives fair and equitable treatment at market."

Viadero added, "I greatly appreciated the efforts of Judith Stewart, U.S. Attorney for the Southern District of Indiana, and Assistant U.S. Attorneys Kathleen Sweeney and Steve Debrota, in the prosecution of this matter of such significance to USDA."

The High Cost Of Cheap Food

The equation between farm prices and food prices has become obsolete as a result vertical monopolies in the food market. With these mega-food corporations making out like bandits on both ends. Shrinking competition in the food industry has enabled agribusiness to pay below wholesale prices for raw materials to farmers while charging over inflated prices to the consumer.

America's farmers and consumers are being gouged.

- Since 1985 cereal prices have soared 90%. Far out pacing any other food product. This prompted Representatives Gejdenson and Schumer to ask the US Department of Justice to look into charges of price collusion between, General Mills, Quaker Oats, Kellogg and Post, whom control over 85% of the cereal market.

- Prices charged for milk have risen more than 15% since 1980 while the prices paid to the American Dairy Farmer has fallen 14%.

- Cattle prices have dropped up to 25% since 1994 with no savings being passed along to the consumer. Over that same time period IBP, ConAgra and Cargill, corporations that control the beef packing market racked up a 100% increase in profit.

- Recently prices paid to the American Hog Farmer dropped by 33% while consumer prices paid for pork at the grocery only dropped by 1%.

"Only four corporations control 84% of the cereals market; 45% of the pork slaughter, 52% of the poultry slaughter market and 66% of the rice market". Al Krebs, Prairie Fire Rural Action

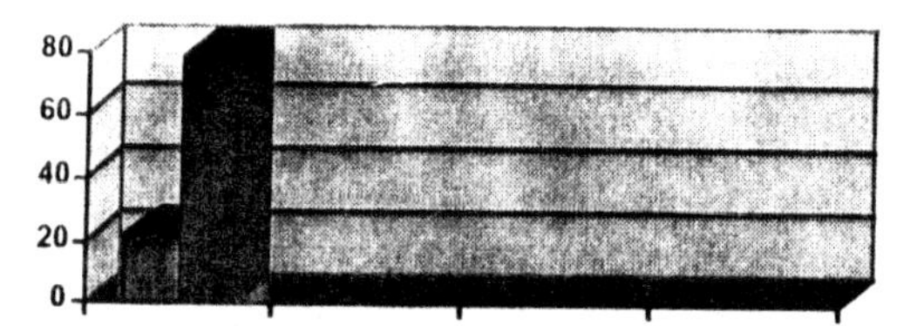

% OF FOOD DOLLAR PAID TO FARMER AND FOOD PROCESSOR

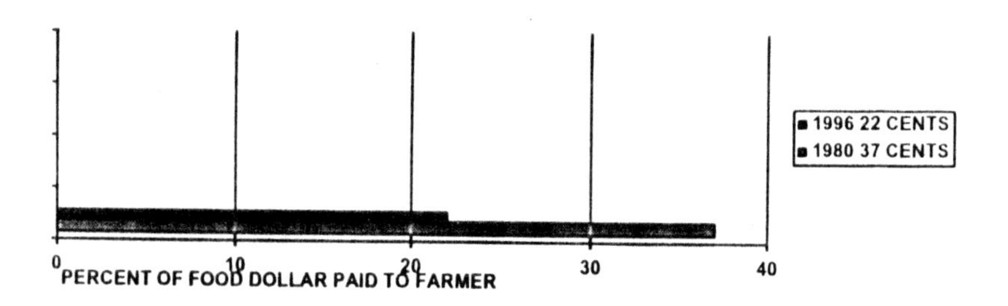

PERCENT OF FOOD DOLLAR PAID TO FARMER

Most Americans would be shocked and disappointed at the amount of their food dollar that goes back to the farmer. Family farmers are not asking consumers to pay more for food. They are consumers to. They just want their share of what sits on the table to be more equitable.

Not only is a lower portion of the food dollar ending up in the farmers pockets but agribusiness has historically paid workers on it's own farms and rural factories near minimum wage. Exasterbating rural poverty. "IBP's use of illegal aliens is a "pristine example" of how agribusiness uses them to drive down wages," stated Pat Buchanan. "That's what I mean when I talk about a two-tiered economy. I'm talking about working people being driven down as though it were a depression at the same time that corporate honchos are making $4.5 million a year. That's not right". Buchanan added "illegal labor, illegal aliens, and real wages of those working people have gone down 40% in 15 years." "This is one of the things that's tearing this country apart."

IBP is the world's largest meatpacker and has a lengthy history of turbulent relations with it's workers.

Corporate greed and mismanagement in Agriculture has also had devastating effects on Americans health. E-coli in Hamburgers, Salmonella in chicken, Mad Cow Disease, Listeria in packaged meats and vegetables are daily subjects on the evening news.

In the book Spoiled The Dangerous Truth About A Food Chain Gone Haywire, the author Nichols Fox reports that figures from the Centers For Disease Control put the number of food poisonings at 81 million cases a year in the US alone.

Hamburger, considered America's favorite food, is particularly susceptible to contamination because the enormous amounts processed at one time. The meat of thousands of cows is mixed together in batches upwards of 40,000 pounds. The batch is then divided and sold and ends up in thousands of kitchens.

In the name of efficiency, one diseased cow in the overcrowded trucks and feedlots can spread disease to thousands. There are so many animals that are moved through so fast a sick animal is usually not detected.

The e-coli virus found in the hamburger at Hudson Food, BeefAmerica and most recently in the 18 tons of meat turned back from South Korea to IBP, Inc. causes bloody diarrhea and kidney failure. Particularly devastating the elderly and small children. In the Hudson case practices of reworking beef left over at the end of each shift to the next days production made it possible for a continuous chain of contamination to occur prompting the largest recall of meat ever, 25 million pounds.

Food contamination is not limited to the meat industry. Mis-handling and importing of vegetables and fruits from other country has become a real problem. Other countries have less to no restrictions on the use of certain pesticides and other chemicals. Airborne disease such as hepatitis have become prevalent on crops coming in from poorer countries. The following is a press release from the USDA prompted by 130 children in Michigan contracted hepatitis in March of 1996.

USDA Release no. 0134.97
USDA ANNOUNCES SUSPENSION OF CALIFORNIA STRABERRY PROCESSOR

WASHINGTON, April 24, 1997 The US Department of Agriculture announced today the suspension of a California fruit and vegetable processing company and its former president form contracting with the Federal Government, charging that it sold to USDA strawberries grown in Mexico for us in the National School Lunch Program. The Federal government purchases only domestic produce for the School Lunch Program, and contractors are required to certify that their produce is grown domestically.

USDA is charging that Andrew and Williamson Sales Company, Inc. of San Diego, California, and its former president, Fred L. Williamson, falsely certified strawberries provided under contract for the School Lunch Program as "100 percent grown and packed in the United States."

The strawberries, processed by A&W in 1996 were linked earlier this month to an outbreak of hepatitis A in Michigan. USDA's Food and Consumer Service issued an order to put on hold and keep from use more than 1 million pounds of the strawberries in 16 states and the District Of Columbia. In an April 1 press release, A&W's parent company, Epitope, Inc. indicated that "A&W inaccurately described some of the strawberries associated with the outbreak as having been grown and processed in the US as require..."

USDA, the Food and Drug Administration, the Centers for Disease Control and Prevention, and the Department of Justice continue to investigate the incident and have not yet determined what charges may be brought against either Williamson or A&W.

Acting Under Secretary for Food, Nutrition and Consumer Services, may Ann Keeffe said both A&W and its former president will be suspended until the investigation and any related legal actions are completed.

"Selling foreign produce to the School Lunch Program under the pretense that is was grown in the US is a very serious violation," Keeffe said. "A&W's parent firm has admitted to that violation by A&W. . We are looking very closely at both A&W and Mr. Williamson to see what charges may be brought against them."

In a letter to both Williamson and A&W, Food and Consume Service Administrator William E. Ludwig said the suspension "will be in effect pending further investigation of the alleged fraud and the completion of any related legal action." Ludwig said the suspension excludes A&W and Williamson from "engaging in most contracts and other transactions Federal...programs nationwide." Both are also barred from acting as a subcontractor under most government contracts.

The Big Squeal Over Pigs

According to University of Missouri Ag Economist Glenn Grimes, demand for pork at the producer level in 1996 was 9% greater than in 1985.

One 80,000 hog corporate factory farm produces as much waste as a city of 180,000 people. In order to contain the waste the corporations build man made lagoons to contain this enormous amount of waste. During heavy rains the lagoons overflow and waste runs off into streams and leak into groundwater contaminating drinking water and endangering people, fish and wildlife.

A spill in North Carolina dumped 25 million gallons of hog sewage over fields and into a nearby 17 mile stretch of river contaminating the water. A recent study by the CDC indicated a link between unusually high rates of miscarriages around an Indiana Hog Farm. Air pollution from the horrible order causes burning of the eyes, nose and throat. In Minnesota a series of air quality tests showed high levels of hydrogen sulfide gas emissions causing headaches, diarrhea, vomiting, and other serious symptoms. Consequently property values for those around these factory have plummeted.

These large factory hog operations can also mistreat the animals in their care. The facilities place thousands of hogs in tight confinement inside large buildings without access to fresh air and sunshine. These hogs never see the light of day. Corporate hog farmers try to persuade communities to let them in by promising jobs which in turn will improve the local community. John Ikerd at the University of Missouri found, for example, large agribusiness contract hog farms producing the same number of hogs as independent producers displace two hog farmers for each new job created by corporate hog farms. This means that a new $5 million investment generates 40-50 new jobs but displaces at least 100 independent producers. A stinky deal.

Big Agribusiness toots it's horn claiming their efficiencies have allowed Americans some of the cheapest food in the world.

Percent of Income Spent On Food

United States	9.3%
United Kingdom	11.5%
Sweden	15.3%
France	16.3%
Japan	19.1%
Italy	25.7%
South Africa	27.5%
India	53.1%

But at what cost does cheap, convenient, fast food really come?

What is not added to the 9.3% figure are costs paid in relation to these food products;

Environmental costs and the money it costs to correct them: Soil compacting, water pollution, erosion, poisoning of wildlife and bulging landfills.
Health Costs: Skyrocketing cancer and other related health problems from dyes, chemicals and hormones in food.
Additional Tax Costs: Billions in subsidies and enforcement costs to regulate the agribusiness sector.

And in the end the consumer pays for overpriced poisoned food. Food is a "basic" necessity. How could something so basic become so complicated? How could something so natural as growing crops and livestock become so un-natural? In other words they are paying to put chemicals on and in our food and in the water used to grow it, polluting our air buy transporting the food nationwide instead of getting it from the local farmer, packaging food in expensive wrappers and filling up landfills. Along the way picking up the cost for exorbitant advertising bills, legislation and enforcement.

Unfortunately, the marketplace and the federal government have placed little value on family farms, rural communities and the health and wallets of the consumer and instead have chosen to subsidize agri-business.
We must ask ourselves if we had these problems when local family farms where growing the majority of our diet. Do we want families or factories growing our food?

The Government and Agriculture

Nationally there are hundreds of programs and services in place to help the farmer and rural community. These programs and services are provided free of charge and are administered by governmental agencies

The government is involved in Agriculture at both the state and federal level. Agriculture policies at all government levels serve to protect, enhance and further farm and ranch product sales. Farm polices have four ostensible goals: keep prices stable, guarantee the supply of food, support farm incomes and ensure the viability of rural communities.

States produce different agricultural products which carry with them different agenda's. States, like the federal government, offer various types of support to farmers and ranchers in the form of technical, marketing and financial assistance.

The department of agriculture at the state level provides much the same services as does the USDA but in particular regards to that states prosperity and economic advancement. Most states agriculture departments consist of divisions that address concerns in the areas of marketing, animal, plant, inspection and business development.

Marketing Divisions: Responsible for developing new marketing opportunities for the states producers and processors. The division develops promotional programs to expand the states farm and processing industries. They can usually assist in developing agricultural marketing plans.

Animal Industry Division: Responsible for protecting the health, welfare and marketability of livestock.

Plant Industry Division: Responsible for pesticide regulations and registration, organic certification, plant nursery licensing, seed dealer registration and matter in reference to crops and plants.

Inspection Division: Responsible for issuing brand inspection to identify ownership of large livestock such as cattle, horses and mules. Responsible for verifying labeling on fertilizers and animal feeds, licensing of agricultural commodity dealer and warehouses, inspection of eggs, slaughter facilities and certification of weighing and measuring devices.

Business Development: Sometimes called an Agricultural Finance Authority. Provides financial assistance to farmer and processors. Administers farm and agricultural programs.

The United States Department Of Agriculture's mission statement is "to ensure the well being of Americans with special emphasis on people engaged in commercial agriculture and sensible management of natural resources, families needing nutritional services, consumers dependent on a safe, affordable food supply and residents of depressed rural areas."

Within the organization of the USDA, the department operates over 200 programs in seven different areas:

Farm And Foreign Agricultural Services
- Farm Service Agency
- Foreign Agricultural Service
- Risk Management

Food, Nutrition and Consumer Services
- Food and Consumer Services

Food Safety
- Food Safety and Inspection Services

Marketing And Regulatory Programs
- Agricultural Marketing Service
- Animal and Plant Health Inspection Service
- Grain Inspection, Packers and Stockyards Administration

Natural Resources And Environment
- Forest Service
- Natural Resources Conservation Service

Research, Education, And Economics
- Agricultural Research Service/National Agricultural Library
- Cooperative State Research, Education and Extension Service
- Economic Research Service
- National Agricultural Statistics Service

Rural Development
- Rural Utilities Service
- Rural Housing Service
- Rural Business Service

In 1993 the USDA employed a nationwide staff of 114,420 and managed a budget of over 68 billion dollars.

The USDA was founded in 1862 by President Abraham Lincoln. He called it the "people's department". In those days 90% of the "people" were farmers who were in need discernible agricultural information. Lincoln established the USDA "to acquire and diffuse among the people of the United States useful information to subjects connected with agriculture in the most general and comprehensive sense of that word".

Since then the USDA's responsibly has grown far beyond just providing agricultural information. Helping farmers and ranchers earn a good living, improving rural American by bringing running water, managing national forests, protecting the soil and water, keeping foreign plant and animal diseases out of the country, increasing agricultural exports and provided meals for 25 million school children everyday are just some of the tasks charged to the USDA.

The majority of a small farmers dealings with the USDA at a local level will be with the Farm Service Agency, Natural Resources Conservation Service, Extension Service and Rural Development.

The Farm Service Agency includes 51 state offices and over 2,500 field offices to serve the nations farmers on a local level. Stabilizing farm income, helping farmers conserve land and water resources, providing credit to new or disadvantaged farmers and ranchers, and helping farm operations recover from the effects of disaster are the missions of the FSA. They administer Commodity Loan Programs, Commodity Purchase Programs, Crop Insurance, Emergency Assistance, Farm Loans and Conservation Programs for American Farmers. Many of the FSA programs are funded through the Commodity Credit Corporation.

The CCC has a $30 billion dollar borrowing authority with the Department Of Treasury. Their relationship with farmers goes back to the 1930's. At that time Congress set up a unique system under which Federal farm programs are administered locally. Eligible local farmers elect a 3 to 5 person county committee, which review county office operations and makes decisions on how to apply the programs. This grassroots approach gives farmers a much needed say in how federal programs affect their communities and their individual operations.

The Natural Resources Conservation Service, formally known as the Soil Conservation Service was instituted in the mid 1930's in response to the Dust Bowl catastrophe. The nation's 3,000 conservation districts, one in almost every county in the US, administer programs and provides technical and financial assistance to farmers and rancher to affirm soil and water conservation. Nearly ¾ of the technical assistance provided assists farmers and ranchers to develop conservation systems suited to their particular land and operation. Programs geared towards the farmer and rancher include: Conservation Farm Option, Conservation of Private Grazing Land Initiative, Conservation Reserve Program, Environmental Quality Incentives Program, Farmland Protection Program, Flood Risk Reduction Program, Forestry Incentives Program, Resource Conservation & Development Program, Stewardship Incentives Program, Watershed Operations Program, Wetlands Reserve Program and the Wildlife Habitat Incentives Program.

The Cooperative State Research, Education and Extension Service provides the agricultural community with a national knowledge based reference system. CREES links the research and education programs of the USDA with more than 130 colleges of agriculture, 59 agricultural experiment stations, 27 colleges of veterinary medicine, 57 cooperative extension services and other schools of science. If you have a question regarding any aspect of agriculture they can answer it usually through you local extension agent.

For questions pertaining to what to raise and how to raise any farm or ranch product in your specific area, your local Cooperative Extension Service is the most reliable source of information. This is a free service to farmers and ranchers and was created by an act of Congress in 1914 to bring the knowledge of Land Grant Colleges to the American Farmer and Rancher.

Your local agent can seek advise from their agricultural experiment stations, plant and animal pathology labs and many other facilities. They can help with decisions on what types of products are best suited for your type of soil, environment, size, water availability, government aid, marketing and financial status. The Service has an extensive library of periodicals and materials published by all the State Cooperative Extension Services and Department Of Agriculture on almost every subject pertaining to farming and ranching usually available free of charge. Your agent can be a great source of information when preparing an Agricultural Business Plan.

The Department Of Rural Development was created in 1994 when rural economic and community development programs had been splintered among various USDA agencies and was created to overcome the problems of the nations rural heartland. More than 53 million people live in rural America and nearly 16% of them fall below poverty level with more than 418,000 households still lacking basic necessities like running water. Rural Development administers Homeownership Loans, Rural Rental Housing Loans, Rental Assistance, Community Facilities Loans, Home Improvement and Repair Loans and Grants among many others. Each year these programs create or preserve tens of thousands of rural jobs and create or improve more than 65,000 units of rural housing.

Chapter 5 – The American Agripreneur – Farmer For The 21st Century

Like the rest of global business agriculture is attempting to adapt to meet the demands of the world. As the worlds population grows the need for food products will more than triple, according to the UN Development Program. The earth's population is growing by 260,000 people each day. Concurrently the supply of farmland is diminishing. Only 1/16th of the earth's surface is tillable. The future will require providing more with less.

At a national level our demographic structure is changing. Consumers needs and desires are changing as the US population is becoming older and becomes more ethnically diverse. Agriculture, intercomparable to the rest of the sciences and technologies, is moving rapidly to keep up with world demand. Consumers are expecting greater convenience, quality and variety. Societies increasing concerns over health, environmental and social issues will also influence production.

American farmers will need to use new technologies to produce something better, at a lower risk with more profit. The acceleration of industrialization in agriculture will mean the agricultural sector will need to seek greater efficiency and lower risks in providing the products that consumers need and desire. New technologies, trends and the decrease of government intervention will usher in a new breed of farm folks.

Adapting to recent dramatic social and technological changes in Agriculture will require a metamorphosis on the part of American Farmers to stay competitive and to stay in business.

Agriculture is changing fast with it's producers obliged to roll with the changes.

- Reduced federal programs: Originating with the 1996 farm bill farmers will start to receive smaller income support payments from government programs. In return they will gain more flexibility but will be assuming more risk. They will need to become better managers and marketers.

- Increased globalization of markets: Food exports will continue to increase which will provide more international opportunities as well as more competition in both domestic and international markets.

- Greater Information systems: The rapidly expanding quantity and quality of information can improve decision-making. Lack of skill or financial resources place many people at a disadvantage.

- New technologies: Agricultural technologies are advancing quickly allowing for acceleration in outputs but are becoming more complex and more difficult to use effectively and profitably.

Technology on the farm has increased at such a rapid rate in just a few short years we have gone from using a horse and home made plow to the use of automated tractors guided by satellites. Farmers are receiving necessary information instantaneously on weather, disease and pestilence. Information that just a decade ago moved so slowly it could have meant the destruction of crops and livestock.

According to a Farm Bureau survey, Computer use on the farm is up to 83.8%, ¾ of farmers used cell phones and 41.9% have a fax. Internet use by farmers is 32.2% up 42% from the previous year.

New farm technologies have dramatically enhanced crop and livestock yields. It grows 2 stalks of corn where only one grew before. The American Farmer is the most productive in the world. Today each US farmer produces enough food and fiber for 129 people.

The development and evolution of farm technology has had a significant effect on production and output. In 1850 it took about 75-90 hours of labor produce 100 bushes of corn with a walking plow, harrow and hand planting. Yields were about 40 bushels per acre. In 1900, 100 bushels of corn were produced with 35-40 hours using a 2-bottom gangplow disk and peg tooth harrow and 2-row planter. Yielding 40 bushels per acre. 1950 saw commercial fertilizer use help increase yield. Corn yields are now 50 bushels per acre. Farmers work 10-14 hours to produce 100 bushels of corn with a tractor, 3-bottom plow, disk, harrow, 4-row planter and 2-row picker. Today only 2 ½ hours and one acre of land are required to produce 100 bushels of corn with farmer using a tractor, 5 bottom plow, 25 foot tandem disk, planter, 25 foot herbicide applicator and 15 foot self propelled combine and trucks.

Today's modern advances in agriculture is the stuff of Sci Fi movies. Worldwide experiments in biotechnology, genetic engineering and robotic farm equipment are headline grabbers. Research and development into agricultural technologies that increase production while not harming the eco system will be required to feed and protect the world for future generations.
Current agricultural advancements include:

- Biotechnolgists and researchers have found a key gene in zucchini and tomatoes and have retooled them to keep production of a natural plant hormone, ethylene, turned off. Ethylene occurs naturally in tomatoes and many other plants. In nature ethylene gets turned on causing fruit to ripen and then rot. Most commercial tomatoes are picked before ripening which shortens the time the fruit can stay on the vine to naturally ripen. Tomatoes with the rebuilt gene can stay on the vine longer. Later, when exposed to ethylene in the warehouse, they soften and turn red.

- Genetic engineering such as cloning and semen sexing are designed to advance meat productivity. Female cows are more valuable to the dairy and male cattle or steers more valuable to beef production. Research is being done at a leading university into semen sexing. They theory is if a sperm carries an X chromosome, the calf will be female. If the sperm is Y it will be male. X; s are larger in size than Y's. The method used to separate them physically is by bathing the sperm in a dye The dye then binds the DNA. The X sperm DNA is 4% greater than Y sperm. The researchers then measure the dye by shining a special laser at the sperm, lighting the dye a camera will detect the light/color and then feed the necessary information into a computer. If the computer measures more light, then there is more DNA. It is then determined that it is an X sperm, which again will yield a female calf. The computers that separates the DNA are still to slow but are advancing.

- Recently new advances in feed allow less pollution in animal waste. An enzyme called Phytase can be added to soybean meal feeds to reduce water pollution around swine, poultry and fish farms. In experiments Phytase has shown that it allows these animals to digest the phytic acid in soybean meal, converting it to the nutrient phosphate. Without the Phytase all the phytic acid ends up in the animals waste. High levels of phosphate feed soil microorganisms that pollute and produce stream-choking algae.

- R&D in non-chemical pest control if the efforts to reduce pollution. Pest extermination above ground is achieved by passing a high intensity electric field of 40 to 50 thousands volts between two electrodes. The current fries the bugs in the field without harming it's foliage or the environment. Pests in the soil will be destroyed by moving electrodes, which penetrate the soil. In much the same way growers disk the soil the electrical exterminator will kill soil born pests with currents passing between each disk as it passes though the field.

- Precision farming: Boosts crop yields and reduces waste by using satellite maps and computers to match seed, fertilizer and crop protector applications to local soil conditions. Global positioning systems and global information services offer satellite technology used to plot field applications. GPS plots information so that every foot of the field can receive customized applications of fertilizer, herbicides, pesticides and irrigation water. The system can report variations of yields, pest management, soil type, soil pH, fertility and more. The system produces a detailed map of the field, revealing a profile. Applying crop inputs on a site by site basis makes for more efficient use of these expensive inputs. Total rates of fertilizer, seed and pesticides will be reduced because they are concentrated where the soils are most productive. The use of this mode of farming will reduce the farmer's production costs and reduce potential farming impacts on the environment.

Current technology has enabled farmers, though research and changes in production practices, to provide Americans with the widest variety of foods ever. Agricultural production is more and more consumer driven. Changing habits in diet have changed the way farmers produce livestock commodities.

Farmers and ranchers are producing meat lower in fat and cholesterol. Retail cuts of meat are 15% leaner. Currently much leaner beef cuts are being produced than were 20 years ago, resulting in 27% less fat. Hogs are bred 50% leaner than 20 years ago.

Recent advancements in biotechnology have found their way into the market place with tastier fruits and vegetables that stay fresh longer and are not damaged by insects.

New Terms And Trends

Most successful small farms and ranches grow different crops and livestock and direct market their products to "niche" markets. These small-scale farmers are taking advantage of changing consumer preference and habits buy utilizing new technologies to produce a specialized product at a lower cost with more profit. This enables the small farm to compete with the big farms. These "agripreneurs" as they are known, are more willing to break out of a traditional crop or livestock product and generally place more effort into the overall management of their farm with a stronger emphasis being placed on sales and marketing.

Recently, Farm Credit Banks, surveyed traditional farmers to find out how they spent their time. The survey found that 95% of time was spent in production. With only 4% spend on marketing and only 1% spent on financial analysis. The study found that farmers that divided their time equally among the 3 activities were much more successful. Today's successful small farmers and ranchers are doing just that. Small farms require more skills to manage the diversification of 4 or 5 different crops and 2 to 3 different types of livestock than a large conventional farm that produces one type of crop or livestock. It also requires the skills to get those products to market in a profitable manner.

The increasing numbers of these small farms and ranches demands a new emphasis is be placed on directing technologies towards making small farms and ranches derive the most profit per acre. Sustainable Agriculture, Organic Farming, Cooperatives and Contract Farming are some new and some old ideas being adapted and applied to the nations small farms.

Sustainable Agriculture

Sustainable agriculture as determined by Congress in the 1990 Farm Bill. Under that law the term sustainable agriculture means an integrated system of plant and animal production practices having a site-specific application that will over the long term:

- Satisfy human and food fiber needs.
- Enhance environmental quality and natural resource base upon which the agricultural economy depends.
- Make the most efficient use of nonrenewable resources and on-farm resources and integrate, where appropriate, natural biological cycles and controls.
- Sustain the economic viability of farm operations.
- Enhance the quality of life for farmers and society as a whole.

The goal of Sustainable Agriculture addresses economic and sociological effects as well as the environmental and scientific aspects of farming and ranching.
The word Sustain means to: to keep in existence; maintain or prolong. Sustainable Agriculture refers to practices utilized by the agricultural communities to provide for it's viability now and for future generations.
Therefore we could say that all new and old practices that benefit and sustain the farm would roll up under the term "Sustainable Agriculture". Practices that meet the needs of the present generation without compromising the needs of future generations.

Sustainable farming practices commonly include:

Crop rotation to mitigate insects, weeds and plant diseases thus reducing contamination by chemicals and reducing soil erosion.

Integrated pest management techniques that reduce the need for pesticides by practices such as scouting, planting times and the use of biological pest controls.

Increased soil and water conservation practices

Increased use of strategic animal and green manure utilization in place of chemicals.

Stresses accommodation to the way natural ecosystems work.

Better management skills, including re-emphasizing the use of labor instead of mechanical intervention and cost cutting practices such as sharing of equipment.

Sustainable Agriculture provides an alternative to high chemical, high machinery, high technology farming. Recent reports have concluded that sustainable agriculture benefits the farmer, environment and rural communities. Researchers from University of Missouri at Columbia, University of Minnesota and the Center for Rural Affairs surveyed a small number of farmers in the northeast Nebraska to examine the effects of sustainable farming on rural communities. The researchers concluded that if all of the farms in the survey had practiced sustainability it would equate to 26% more people residing in the community, therefore causing an 80% rise in the property tax base.

Contract Farming

Contract farming is the most common way of selling farm products today. Over 99% of the nation's poultry growers are farming under contract. Contract Farming is farming in the traditional sense but backwards.

Typically a farmer produced what he hoped would give him the most pounds or bushels then offered it for sale for whatever he could get for it. Under contract Farming the farmer "contracts" with a processor before hand to raise crops or livestock to the processors specifications. The farmer or producer provides the facilities, utilities and labor. The processor or investor provides the animals, feed, medicine and management direction. As compensation the joint venture can be based on a per head or bushel agreement or on a percentage basis. Per head is the least risky for the farmer but has the least amount of profit. Under percentage the risk is shared but then so is the reward if the market is good.

Contract farming in its inception was set up as a good idea but recently has come under tremendous scrutiny and criticism of processors for under paying farmers leaving them in break-even financial conditions and subsistent lifestyles.

Intensive and Rotational Grazing

There are a number of very successful grazing methods that can increase your stocking or per head rate 2 to 3 times. Many of these methods employ dividing up your farm into cells and paddocks and rotating your livestock frequently Proper fencing and watering systems determine how much livestock per acre can be attained. The purpose behind these grazing methods is to utilize land to it's utmost capabilities thus increasing profit by reducing or eliminating the need to supplemental feed. These types of grazing systems can reduce feed costs and allow more animals on a smaller size farm. Grazing systems are highly labor intensive, requiring the movement of animals every couple of days from paddock to paddock. Information on a number of grazing methods can be obtained through the local extension agent.

Risk Management

Risk has always been a part of farming. Protecting the investment a farmer puts into crops and livestock through careful cash flow planning is more important than ever before as the USDA slowly backs out of price supports.

Risk Management is the practice of carefully contemplating actions in which long-term economics of those actions such debt assumption for land and machinery purchases and management decisions such as participation in programs like the CRP affect the farms viability.

Risk management can include: Financial practices such as negotiating flexible financial contracts or deferring payments as farm income conditions vary. Weighing the advantages of diversification strategies of crops, livestock, and production locations to stabilize farm income and utilizing various forms of financial protection such as crop insurance and revenue insurance programs.

Alternative Agriculture

The terms Alternative Agriculture encompasses an array of "non-traditional" enterprises. They include: Non-traditional livestock, crops and other farm products, unconventional production systems such as aquaculture or organic farming, entrepreneurial marketing strategies, and end product production.
The following review of 20 Alternative Agriculture Products offer a "higher profit per acre" and are non-traditional which would categorize them as Alternative Agriculture.

ALLIGATOR Alligators have been hunted as early as the 1890's for their skins and meat. The first alligator farm was established in Florida in 1891 and has sustained itself as a viable industry since. However recently renewed interest ir e alligator industry has grown as the new aquaculture farm movement has taken off. In contradiction to popular belief, Alligator farming does not adversely affect the environment nor does it require a lot of land and water. For information on Alligator Farming contact one of the five regional USDA Aquaculture Centers listed under Aquaculture.

AQUACULTURE Aquaculture or "fish farming" is one of the fasting growing ranching alternatives. In the last two decades the catfish industry has grown about 20% annually. In 1996 500 million pounds of Catfish were sold. The public's demand for health food and fish that is safe from the pollutants dumped into our rivers and oceans has spurred the rapid growth in this industry. Local, state and federal government regulate this industry. All appropriate agencies should be contacted before beginning any aquaculture project. Intensive research should also be undertaken. More than 100 different species of fish can be farmed. Your local agencies should be able to tell you which fish are best suited for your particular area. Then, if not familiar with fish farming, studying the biology, chemistry, management and economics of producing the fish should be undertaken by the potential rancher. Site location and availability of water should also be determined as well as determining the financial resources it will take to start such a project. One source of financial funding can be the USDA. The Farm Service Agency of the USDA does accept applications for small-scale aquaculture enterprises. For more information on Aquaculture Enterprises you can contact on of the 5 regional USDA Aquaculture Centers: NCRAC, Room 13 Natural Resources Building, Michigan State University, East Lansing, MI 48824. SRAC, Delta Branch Experiment Station, Stoneville, MS 38776. TSRAC, The Oceanic Institute, Makapuu Point, Waimanalo, HI 96795. NRAC University Of Rhode Island, Kingston, RI 02881. WRAC School of Fisheries, WH-10, University of Washington, Seattle, WA 98195.

BAMBOOFor thousands of year's bamboo has provided food, shelter and building material for most of Asia. In the North America in recent years bamboo has been rediscovered as a landscaping plant and is in high demand. Because of it's fast growth into a very large plant it is being used for shading instead of trees. It does not take years to mature like a tree would. Bamboo can be used as a hedging also and keeps it's green color throughout the winter. It can also be placed in pots and makes a dramatic houseplant. Prices can range from $100 to $200 a plant. A misconception of Bamboo is that it must be raised in a tropical condition. Many varieties come from Japan and West China where conditions can be similar to the US East Coast. For more information: Homestead Growers Bookshelf, PO Box 2010, Port Townsend, WA 98368.

BEEF Many first time ranchers shy away from the cattle industry because it is thought that in order to raise cattle you will need a large sprawling ranch and a large sprawling wallet to get started. However, small-scale cattle production can be profitable. Today's new technologies such as electric fence and better, intensive grazing management has proven that more cattle can be successfully produced on smaller amounts of land. Rotationally grazing animals by moving them into temporary paddocks, secured by an electric movable fence, can reduce overgrazing and reliance on outside feed sources. Also non-traditional marketing means such as selling cattle direct to the consumer as leaner, organic, medication free animals can bring as much as $200 to $300 a head more than running them through the sale barn. For more information: The book "Salad Bar Beef", by Joel Salatin, QYLM, 152-RS3 Maple Lane, Harriman, TN 37748. "Intensive Grazing Management: Forage, Animals and Profits" by Burt Smith, Ping Sun Leung and George Love.

BEEKEEPING Over 200,000 beekeepers sustain over 3 million colonies in the United States. Each hive can produce over 100# of honey a year. The production of honey, beeswax, pollination rental and sales of queen bees can be successfully propagated on just a few acres. However, before entering this industry one must gain knowledge on bee management, biology and flora. For more information: American Bee Journal, 51 South 2nd St., Hamilton, IL 62341. The local library should have a variety of beekeeping books

BERRIES Few crops are as well suited for small acreage as Berries. On just 10 acres you can produce over 16,000 pounds of Blueberries. If sold at market for only $1.00 a pound. That's $16,000 a year. The more exotic the Berry. The more the profit. Brambleberries can be sold for as much as $5.00 per half-pint in urban areas. Yielding 5,000 pounds and acre a gross return would be about $13,000 an acre or $130,000 on your 10 acres. Because of their delicate nature berry production can be labor intensive. Each acre may require about 10 pickers. A side value added line can also prove to be profitable. Homemade jellies, syrup, candy, juice can be successfully marketed to "Yuppie Markets". More information: The North American Bramble Growers Association, Richard Fagan, Executive Secretary, Rt. 2, Box 539, Cumberland, MD 21502. Your local Agriculture Extension Agency.

BISON Buffalo on the hoof can sell for up to 100% more than cattle. Especially when selling breading stock. They are leaner than beef and are recommended as an alternative to beef by the American Heart Association and National Cancer Institute, which has driven the demand way up. In addition to fetching a higher price, Buffalo are a heartier animal, have less illness than cattle requiring less vet bills and expense of death loss. Buffalo also have a longer reproductive lives, about 3 to 4 times that of cattle. Buffalo will adapt to land to poor for cattle to graze on. All around are more profitable and require less attention than cattle production. For more information: American Buffalo Association, PO Box 16660, Denver, Colorado, 80216. (303)222-2833

CHRISTMAS TREES Every year the constant grumbling about the high prices of Christmas Trees leaves the Tree Farmers smiling all the way to the bank I'm sure. In some urban areas trees selling for over $100.00. Whether it be choose and cut or selling on lots, Christmas Tree Farming can be profitable. Well drained soil on about 5 to 40 acres is sufficient to obtain a gross return of about $10,000 an acre with few out of pocket expenses. For more information: The National Christmas Tree Association, 611 East Wells Street, Milwaukee, WI 53202, (414)276-6410.

ESCARGOT Actually snails. The US imports millions of the little creatures every year for our culinary consumption. Anyone that has ever bought Escargot as an appetizer knows that you get about six on a plate for up to $15.00 depending on the restaurant. Snails are high in nutrients with the least amount of calories to comparable foods. Escargot probably has the lowest start up costs of any form of alternative agriculture. You can raise about 15,000 snails in about 800 square feet. State regulations should be checked before ordering or raising snails. For more information: The Snail Club Of America, 187 North Duke Ave., Fresno, CA 93727

FLOWERS According to the USDA, today the fastest growing segment of commercial agriculture is flower production. Demand is growing by 10% each year. Highly enjoyable and profitable, in the last decade traditional flowers, wild flowers, edible flowers and dried flowers have found renewed popularity with the public. Value added products such as the old Victorian practice of candying flowers can sell for $45 per hundred. For more information: The Association Of Specialty Cut Flower Growers, 155 Elm St., Oberln, OH 44074.

FOLIAGE PLANTS From the 1970's to the 1980's the number of acres and values of foliage plants, especially tropical plants, has grown and increased sales by 834%. From 3,000 square feet a handsome profit of $10,000 to $20,000 per year can be attained. Foliage plant production can range from a backyard size to 80 acres. Some produce their foliage outside but most producers grow in greenhouses. Special attention to overhead such as heat costs can determine the viability of the project. For more information: "Foliage Plant Production"., by Jasper N. Joiner, Prentice-Hall, Englewood Cliffs, NJ 07632.

FUR PRODUCTION Animals such as mink, lynx, fox, chinchilla and rabbit are well suited for small acreage. On the downside it can be tough to get started in this industry. A breeding pair of mink can set you back thousands. Then there is the cost of housing and caging. On the up side the market for this fur is w[illegible]d wide. The key is to start small. Most large breeders started with only a few animals. Pelt sales can range anywhere from $16.00 per pelt for a rabbit up to hundreds of dollars for the more exotic breeds.

GAMEBIRDS The game bird industry has grown tremendously over the past two decades. Pheasant, wild turkey, partridges and ducks are under increasing demand in up scale meat sales and by hunters. The number of these birds has decreased while the number of licensed hunting preserves is on the increase. Many hunters pay to belong to these preserves in order to have a successful shoot. Dressed pheasants can sell for $3.75 a pound and can sell for up to $12.00 each for live birds headed for the hunting preserve. Around 2000 pheasants and 4000 quail/partridge can be produced on 1 acre. For more information: American Pheasant and Waterfowl Association, Route 1, Granton, Wisconsin, 54436. (715)238-7291

GAME MEATS Like game birds, elk and deer ranching is a fast growing industry. Both for meat sales to the upscale, health conscious, baby boomers and to hunting preserves. Hunting preserves prefer ranch grown elk and deer because they are accustomed to being fenced in. A large wild elk crashing through a fence either killing or maiming itself is expensive for the preserve. Sales of live trophy bulls to a game preserve can bring up to $10,000 each. Prime cuts of elk meat sold to restaurants or upscale stores can go for as high as $10.00 a pound. Deer for $6.00 a pound. Expect to invest a good amount of money though. Breeding Stock, 8 foot high fence and land. For more information: Exotic Wildlife Association, 1811-A Junction Highway, Kerrville, Texas 78028. (512)895-4288. Stagline, New Zealand Deer Farmers Association Inc., 4th Floor, Agriculture House, Johnston Street, PO Box 2678, Wellington, New Zealand. North American Deer Farmers Association, Lucky Star Ranch, Box 273, Chaumont, New York, 13622. (315)649-5519.

GARLIC Garlic has been around for thousands of years and has been utilized for medicinal purposes, cooking and as an organic pesticide. It is a member of the same family as onions and is almost foolproof to raise, producing at 15,000 pounds per acre. Sales of Garlic can range from $5.00 a pound to $15.00. For more information: Homestead Grower's Bookshelf, PO Box 2010, Port Townsend, WA 98368

GINSENG Over 50 million dollars in Ginseng root is exported from the United States annually. Ginseng can gross about $44,000 to $70,000 an acre. However it can be labor intensive and require patience. Time to harvest can range from 4 years to 10 years depending on conditions. For more information: American Ginseng: Green Gold, W. Scott Persons, Tuckasegee Valley Ginseng, Box 236G, Tuckasegee, NC 28783.

GOATS Goats can be a dual purpose animal. It was surprising to me just how many people are unaware that both Cashmere and Angora, fibers used to make fine sweaters and suits, come from the lowly goat. Goat meat is eaten by 2/3 of the worlds population. And that most of your high quality cheeses come from goats milk. Goats are easy to raise, easy to find good breeding stock and are an enjoyable animal to have around. A goat owner once told me that goats are the only heard animal they know of that live to play. As a goat owner I find their antics quite amusing. Goats are good producers having single to triple births. Goats are a hardy animal. They are browsers and forage on weeds, leaves and just about anything they can find. For my goats the stickers off the tumble weeds seem to be a culinary delight for them. They are also more disease resistant than most domesticated herd animals. For more information: American Dairy Goat Association, American Meat Goat Association (513)384-2829, Cashmere America, PO Box 1126, Sonora, TX 78950 (915)387-8052.

HORSE BREEDING/BOARDING $850.00 a month to board one horse. Yes this is what some horse borders located just outside of large metropolitan cities can charge for their services. This price would most likely include feeding, grooming and some training for the horse. Adequate stalling, fencing, pasture, feed and water supplies and manure management must also be a consideration. But if you have a barn that will stall 10 horses and kept them filled on a consistent basis, yearly gross receipts could total $102,000. For more information: Horsekeeping on a Small Acreage by Pownal, Garden Way Publishing, Sotrey Communications, 1990.

KIWIS Unlike Apple or Orange trees which can take years to give a good harvest. A kiwi vine will bear in 2 years, maturing in 4. Kiwifruit prefer sandy soil but are highly adaptable to any soil . higher in vitamin C than an orange (about twice the amount) they also contain iron and other minerals that oranges do not. Kiwi's are also higher in vitamin E than an avocado but with only 60% of the calories. With it's high nutritional value the Kiwi is said to resemble more of a vegetable than a fruit. Kiwis are also believed by some in the scientific community to contain an enzyme that is being studied as a retardant to the human aging process. 5 acres of Kiwi's can produce about 100,000 pounds averaging about 35 cents a pound being marketed as an specialty organic product. For more information: Contract your local Extension Agent.

MUSHROOMS Mushrooms are an ideal crop because of the low start up costs and can be produced almost anywhere. In fifty pounds of straw you can produce about fifty pounds of Oyster Mushrooms. In a 100 square foot shed or outbuilding in the backyard it is possible to produce about 3000 pounds. At $6.00 a pound that's 18,000 from a very small investment. A mushroom being neither a animal nor plant contains the properties of both. For more information: Mushrooms in the Garden by Hellmut Steineck, Mad River Press, ISBN 3-8001-6122-2. Homestead Grower's Bookshelf, PO Box 2010, Port Townsend, WA 98368.

PET BIRDS Limited on space many urbanites who long for the companionship of a dog or cat have sought out alternative forms of friendship from pet birds. More than 6% of American Homes have a pet bird. City dwelling pet owners that cannot own a traditional pet because the rules and regulations of apartment or condo living are usually allowed pets that can be contained such as birds or fish. Many agriculturists have overlooked this industry. Cocktails, Love Birds and Budgies can be raised with a nominal investment and some research. Successful breeders can sell the birds for up to $150.00 each. Since this endeavor is uncharted territory for most farmers and ranchers, starting out small and gaining knowledge as you go is highly recommended by pet bird breeders.

SPECIALITY VEGTABLES AND FRUITS The growing demand for specialty vegetables and fruits such as ethnic or unusual foods has the major food stores across America looking to the small scale growers to supply these hot products. The growing ethnic population in the United States coupled with the health conscious "yuppie" group is causing this industry to rapidly expand. Oriental, Hispanic, Native American, Miniature or Gourmet vegetables and fruits can be sold to major chains, restaurants and upscale markets at prices far higher than those of traditional produce. An estimated gross return on 1 acre of Chinese Cabbage can be from $9,000 to $12,000. For more information: Windows Of Opportunity: The Market For Specialties and Organics by Nancy Lee Bentley, PO Box 62, Cabery, IL 60919. Contact your local Extension Agent to find out what crops will grow in your area.

VALUE ADDED AGRICULUTRAL PRODUCTS Goat Sausage, Blueberry Jelly, Homemade Salsa, Cheese are all examples of value added agricultural products. Many farmers and ranchers are cutting out the middleman, adding a little marketing savvy are taking their products directly from harvest to end products and reaping the profits. Pickling and marketing as a gourmet food product can bring up to 6 times more profit from the same produce. People just don't have time to pickle, make jelly, bake but still yearn for that home made taste and are willing to pay for it. Many farmers and ranchers tired of the ups and downs of livestock and produce prices have found that producing value added products stabilizes their income by fixing a cost to what they produce.

Agricultural Diversification

$100,000 from 25 acres? That's what Booker T Whatley's, retired professor of horticulture from the Tuskegee Institute claims in his book of the same name.

His plan is simple in theory. Produce at least 10 different high val·· · products that will give you a year round cash flow instead of traditional annual crops such as corn or wheat. By high value, he means crops that will gross a maximum of $3,000 an acre. He advises to select crops that mature at different times so they don't compete for harvest labor. "By diversifying you greatly reduce risk". If you have 10 crops each, contributing about 10% of gross income, even if 1 or 2 crops fail in any year, losses wont exceed 10-20%.

Traditionally farmers only grow one or two crops. Adversities in whether or markets can become financially devastating to a years income. Larry Statler director of the Rural Diversified Enterprise Center agrees "Farm businesses, large and small alike need multiple profit centers".

Organic Farming

Organic Farming is a profitable and environmentally sound farming practice that integrates technology with proven ecological sciences. Beneficial Insects, Composting, Organic Pest Management and non-medicated feeds are all natural ways to grow crops and livestock. These systems are environmentally sensitive, cost effective and resource conserving. This system eliminates the need for chemical fertilizers, insecticides and livestock feed additives all of which are expensive and harmful to the environment, the livestock, the farmer and consumer. This farming philosophy can bring many benefits to farmers and ranchers in the form of less cost but more profit by being able to make his products "Certified Organic".

Since World War II chemical sensitivity from the more than 50,000 pesticides and synthetic chemicals has become such a problem that IRS will make allowances for organic foods as a legitimate tax deduction

The Food Marketing Institute, in a recent survey found that ¾ of all shoppers have a major concern about chemical residues in food. The National Academy of Sciences recently published "Regulating Pesticides in Food" in which the authors point out that the American food supply system may contain sufficient levels of residues of various agricultural chemicals to be of some long term health concern.

Organic food now accounts for 1% of the nation's grocery bill making it a 3 billion-dollar industry. Seeing growers charging 20-50% more for these products. Organic cuts of beef can go for as high as $9 a pound, bacon $5 a pound, Grapes $3 and Apples $2 per pound.

Conservation Farming

Conservation farming is the use of conservation methods in which to reduce soil erosion and retain soil moisture. Methods used by a conservation farmer include: Conservation tillage, crop rotation, contour farming, stripcropping, terraces, diversion and grass waterways. In 1996 these methods, which promote less soil disturbance, were applied to over 109 million acres, a 6 million-acre increase from the year before or 37% of All-American farmland.

Additionally 77 million acres utilized "reduced tillage practices". Conservation farming leaves crop remnants like leaves and stalks on 30% or more of the soil surface after harvest. The leaves and stalks turn into organic matter, which is then tilled into the soil. This practice can reduce soil erosion by as much as 90%. Conservation Farming also benefits farmers in the area of lower costs. Such systems reduce the number of times the producer has to make through the field for planting and cultivating. These methods save farmers time wear on machinery and the need for expensive inputs while building soil productivity.

Internet Use in Farming and Ranching

In a recent study of the rural population 97.9% of respondents stated that the Internet us very useful for gathering farm business information and 70% strongly agreed on the Internet for it's importance for rural use.

Any time night or day, 24 hours a day, 7 days a week any farmer or rancher can access the USDA's computer library, individual states Divisions of Agriculture and Extension Office Libraries. Major agricultural magazines and newspapers and farm chat groups offer aid and support. Literally millions of articles, papers, books, periodicals and advise all relating to agriculture right at your fingertips any time. The Internet is turning out to be a valuable asset to the sometimes-isolated farmer or rancher. Up to the minute whether and commodity pricing keep Farmers and Ranchers better informed to make quick and concise management decisions.

The Internet was created 20 years ago by the Department Of Defense as a way of securing communication in the event of nuclear war. In the early 1990's the government turned over many of the high-speed data links to commercial communications networks. These communications networks also known as services, such as America Online or BM, will send you a free copy of their software for signing up with them. You must pay a monthly access fee which generally runs about $20 a month for unlimited use. The software and service is very user friendly and can be a tremendous asset to farm management.

E-mail a question for an expert at the USDA in Washington or your Local Extension Agent. Get in a chat group with other ranchers that breed your livestock both here and in Australia. Receive daily updates from the USDA on commodity prices and farm policy. Receive an up to the minute weather report. Advertise your agricultural products worldwide on your very own web-site. Download farm loan applications and paperwork from the USDA.
Here are the e-mail addresses for the more popular agricultural sites:

Successful Farming Online	www.agriculture.com
Agri Alternative Online	www.agricalt.com
Top Producer	www.topprodcer.com
High Plain Journal	www.hpj.com
USDA Homepage	www.usda.gov
Farm Service Agency	www.fsa.usda.gov
Natural Resources Conservation Service	www.ncg.nrcs.usda.gov
Agricultural Research Service	www.ars.usda.gov
National Agricultural Library	www.nalusda.gov
American Farm Bureau	www.fb.com
Cooperative State Research Education and Extension	www.reeusda.gov

Modern Marketing Techniques For Agricultural Products

Cooperatives

A cooperative is an organization formed for the purpose of producing and marketing goods or products, owned collectively by members, who share in their benefit. "For more than 100 years, cooperatives have played a vital role in helping farmers improve their ability to market and process their crops and livestock and to secure farm supplies and other services at reasonable costs", Long Thompson, Agriculture Under Secretary.

In 1996 farmer owned cooperatives set new business volume and net income records for the 2nd consecutive year. America's 3884 farmer cooperatives reported record net business volume of $106 billion for 1996, a 13% increase over the previous year. Cooperatives realized a record net in 1996 of 2.37 billion, up from 2.33 billion in 1995. This is due in part to Cooperatives getting more and more sophisticated - graduating from seed buying clubs to globally competitive, vertically integrated exporters that band small farmers together in effect to mimic successful large operatio- These kinds of cooperatives attain higher prices for members by successfully competing with similar corporate owned entities for market share.

To compete, farmer owned and operated processing cooperatives are employing the following strategies: focusing on retail and consumer demands, providing quality products, capturing newly emerging markets, providing unique products or diversifying risk by establishing a strategic relationship with other processors in the market.

Across the nation family farmers, who are often unable to afford the initial capital necessary to open a processing facility individually, are joining together to form processing and markeiing cooperatives. These new farmer owned and run enterprises differ greatly from manv of today's corporate owned cooperatives.

These new processing cooperatives are owned and operated by active farmers and ranchers who become members by purchasing shares in the operation and supplying additional start up capital to fund new equipment, research and marketing costs. Memberships and stock transfers are usually be review by the board of directors who make sure that new members are actively engaged in farming.

Some of these farmer owned cooperatives, rather than producing raw cheap raw commodities, are forming value-added cooperatives and then marketing them directly to consumers. These cooperatives are successfully marketing their products to consumers demanding safe quality, locally produced food. These farmer owned cooperatives are milling their own grain, processing their own livestock and producing their own brands of pasta.
Example: You cant make a living selling corn for $1.90 a bushel but if you mill it yourself you can net $4.00 to $5.00 per bushel.

Government financial and technical assistance is available to any producers that want to start a coop. One of the sources utilized by cooperatives is the Agriculture Cooperative Service, a division of the USDA in Washington, DC. They have material (sample by-laws, incorporation, membership forms, etc.) which can be sent to any interested organization. They will also follow up with visits to help in organizational development.

Direct Marketing

Farmers that direct market their products can derive up to 80% of their income dollars from marketing their own products in comparison to only 20% from regular production means.

Direct marketing allows the farmer or rancher to respond quickly to consumer needs and market locally where they can maintain more control over product distribution, pricing and quality. When surveyed food customers seek quality first, freshness, ripeness and flavor and are willing to pay more for these qualities. When Direct Marketing stressing your quality, using words like fresh, local, lean, organic can demand a higher price for a farmers products. A realization that there are factors other than price that motivate purchases maintain the quality of your produce.

Farmers Markets

Farmers markets are growing in numbers and profits. These markets often provide a lifeline of income to smaller operations by allowing farmers to sell directly to consumers, keeping the full profit of their farm products for themselves. In California the number of small farms is actually rising and folks there attribute the climb to the popularity of farmers markets.

Marketing To Wholesalers

When approaching a food wholesaler you will need to: determine their packaging requirements, determine the quantity of product they move, determine what post harvest care is required. If you can meet these requirements you will need to present them with a business and scheduling plan and then an agreement on pricing strategy.

Pick Your Own

A pick your own can be very profitable if set up correctly. These types of farms offer an "event" for a family not just a trip to the grocery store. First you will need to determine the number of people in your 30-mile radius. Do you have enough of a market that can support your farming enterprise? Here are some approximate guidelines for planting a pick you own:
Plant one acre of strawberries for every 8,000 persons;
Plant one acre of raspberries for every 16,000 persons;
Plant one acre of asparagus for every 14,000 persons..

Marketing to Restaurants

Many restaurants, searching for quality and freshness, have become increasingly open to dealing directly with the farmer. Most are looking for quality and can turn a nice profit by advertising locally and organically grown dishes. Products that are not normally available through mass distributors such as pheasant or ostrich meat can offer them variety.
Market research into local restaurants to determine their needs starts with:

- Being familiar with the local restaurant scene
- Visiting restaurant associations and trade food shows
- Do Yellow pages or dining guide surveys
- Read food magazines for trends

Home Processing

Home made sausage, jellies and cheese are all end products that can be processed from farm products in a home environment. However home food processing can fall under many regulations:

1. Check with county health board for standards
2. Check with FDA for labeling requirement
3. Check with USDA if processing meat
4. Have a laboratory check for safety of your product

Marketing to Specialty Stores

Wild Oats and Alfalfa's Markets in Colorado are specialty supermarkets that cater to organic food buyers and usually deal directly with local farmers. Alfalfa's recently reported sales of 16 million a year. These specialty chains generally charge about 20% more for organic.

Mail order

With improved packaging and over night delivery service to virtually anywhere in the country mail distribution of meat and produce is becoming a hot market. Each year at Christmas my sister, who lives in Alaska, sends Alaskan king crab to the family via next day service. Mail order products generate 10 billion pieces of mail at a cost of more than

2.5 billion according to the Direct Marketing Association Inc. Mail order has become a 150 billion-dollar a year industry with Agricultural Products emerging as a strong addition to this already popular way of shopping.

Chapter 6 - Starting A Small Farm

Starting a small farm is the circumstance in which most new or beginning farmers find themselves. The majority of new farmers under the age of 35 have a net worth of less than $100,000. It is estimated to buy an established full-scale farm enterprise would require an investment of $500,000. Far and away out of reach for the young farmers budget. The prices of land, equipment and livestock remain at an all time high with collateral requirements for agricultural enterprises being some of the most stringent in the commercial banking sector. Therefore the majority of beginning farmers or ranchers start from the ground up. They establish a small part time farm, continuing to work a full time job, until a level of profit is reached from the farm business to sustain the family and the farm.

Starting a small farm from scratch leaves many options to be investigated and weighed. Numerous decisions to be made and details to be overseen. Your new small farm will require a great deal of commitment, innovation, flexibility and management skills. An inventory of assets both personal and financial can ascertain a better perspective on the direction of your farm or ranch enterprise.

Do you have the skills, knowledge, financial resources and determination to make a life on the farm? The decision to farm is a drastic one that is filled with self-realization and soul searching. It is in essence bucking societal pressures to conform to an urban or suburban structure and can be perceived by others as a rebellious or even "crazy" feat. Many people probably wont understand your desire to farm and ranch. Expect a lot of flack over your decision from family and friends. This reproach usually come from ignorance and general lack of knowledge over the state of agriculture. In the past the media has focused on the farm crisis and families losing their farms. Rightly so, but they do not report on the numerous small farms that are profitable and sustaining families in a comfortable living. If you have the resolve best bet is to turn a deaf ear or if you are truly cut out for farming your independent spirit will make itself evident when you urge your critics to mind their own business.

A general preference for a farming lifestyle and rural living is essential. If you can adapt and thrive in isolation and are of an independent, brave character you have the basic clay in which to form farm life. A natural fondness for the business A want to nurture the soil, crops and animals and an enjoyment of the outdoors are characteristics of a true farmer. This is also an essential trait for the rest of your family. Farming is a family business and lifestyle.

My belief is that in the next 100 years there will be a great decline in "acreage" ownership. Eventually land of any significant size with be irreplaceable. Suburbia is popping up everywhere, especially in rural areas. Builders have caught on to the consumer preference for rural living and are buying up farms, subdividing and selling tract homes in rural America at a tremendous rate. Thus squeezing out the agricultural faction. The utter number of U.S. residents will demand more property of less size. Like many other farmers I am building my place to leave to my children and therefore the decision to start the ranch was a family decision for the benefit of all.

There are strong arguments in favor of starting an agricultural enterprise as a hobby and a part or full time small business.

- The long-term gain in investment in "acreage" property builds greater net worth. Wil Rogers stated "Buy land they aint makin anymore of the stuff". The preference for rural living has driven demand up and supply down. This trend is predicted to stay strong in the future. The value of acreage property is expected to at least hold it's value if not increase significantly.
- Livestock and crops can be a good investment. As a savings account they generally perform better than CD's and stocks and yield a higher amount. They are also good collateral for loans as the are very liquid.
- Generally agricultural equipment holds it's value more so than many other types of vehicles and machinery. Equipment is also easily liquidated and can be held as collateral for a loan.
- Taxable income can be greatly reduced by the write off of qualified expenses from a "farm" enterprise.
- Farming can provide a great second income that the whole family can share in.
- Current trends in government assistance both at the federal and state level through low interest loans or grants makes it easier to obtain the necessary financial backing for an agricultural venture.
- Reduction in the family's food expenses and an overall health benefit. Farm families can sustain themselves by growing their own produce and meat products in a safe, organic manner. Experts say that for every $20 spent on gardening supplies, $200 is saved on the family's grocery bill.
- Many individuals enjoy farming and ranching as a rewarding hobby or career. The social activities and community events surrounding farming can be fun and satisfying.
- Farming provides a good environment in which to raise children.

Do you have the skills and knowledge to own a farm? If you have the willingness to learn the answer is yes. Farming and Ranching is not hard it is just largely un-taught. In the skills category the farmer's resume would certainly list the following experience and abilities: meteorologist, biologist, chemist, mechanic, handyman, veterinarian, manager, accountant, salesman and financial analyst. All of which would require a number of college degrees. Most farmers became qualified in these areas by on-the-job training and an ability to learn. In addition the farmer has an uncompromising determination to the permanence of his or her farm. What ever is necessary to sustain the farm is what he or she will do regardless of the leaning curve required to perform the tasks.

The skills and knowledge one accuries is largely through life's experiences. Obtained through other people, their surroundings and events in their lives. Willingness, enthusiasm and determination are all character strengths that lead to a high rate of comprehension in disseminating information. Without these character strengths the ability learn new things would be greatly diminished and farming is no different.

The basic skills of farming necessitate the following essential character traits:

Bravery - The courage to take deliberate risks. The ability to step up to the plate confidentially and overcome adversities such as the ups and downs in the weather, the onset of disease and pestilence, market politics, consumer preferences that can swing temporarily or permanently and geographic changes such as the expansion of local communities that may place constraints on your ability to farm as you please. As in the rest of the economic world, farming can by risky business, but now a days so is working for someone else. The ability to stand behind your decision to farm or ranch contrary what's popular or "trendy". Strength of character and moral conviction to respect the environment and the livestock and land placed in your care.

Optimism - To focus on the positive aspects of farming and to press on when things get a little rough. In farming there is always next year. Remember if there were no good years in farming, there would be no farmers. To have a fun, adventurous outlook towards the day to day operations. Farming and Ranching are never boring and are always transitional.

Determination- A resolve to permanence no matter the setback. I don't mind telling you that the ranch has brought me to tears more than once. However, the lasting commitment to my vision is unwavering. In farming you are not only committing to a business but a lasting lifestyle. One can experience a love - hate relationship with the farm; occasionally feeling it has betrayed you in some way. Most farmers still would not trade it for any other way of life. It can instill in you a passion for the business and devotion towards the self-discipline and hard work that is required to sustain a viable and profitable business and lifestyle.

The Small Farm Inventory

What crops or livestock to raise? What type of property to select? With what types of facilities? What outlets are there for your products? These and many more questions should be examined and carefully planned out before the initial investment of time and money in a farm/ranch enterprise.

Products

If you have not already purchased your small farm or mini-ranch it would be a prudent move to decide what types of products you want to raise first. To build the necessary facilities or renovate an existing facility to meet the needs of your crop or livestock production is much more expensive than locating a farm or ranch that is already equipped with these facilities.

The initial idea for the type of product you want to raise can have a lot to do with personal considerations. Preference, time and financial concerns may play•a role in the products you choose. Some ideas to consider when choosing your farm product:

- Hobbies. If you prefer hunting, you may want to consider a game animal such as deer or pheasant. If you enjoy rodeos you may want to raise horses. If you stick with something you truly enjoy chances are you will put more of an effort into the success of the product.
- Select a mix of products that will give you year round income. Diversity stabilizes income and minimizes risk. Strategize ways to produce revenue. When one product goes out of season what will be it's replacement. Or retain a product that produces year round. Pigs and goats deliver their young year round other

livestock does not. To farm all year you will be utilizing the most from your resources.

- Determine which enterprises are best suited for the land and climate. Most fruits and vegetables need irrigation or heavy amounts of rain and moisture. Other crops such as hay can withstand dry land conditions. Animals, except for some foul, do well in just about any climate as long as they have shelter. Select an enterprise that will do well in its natural environment.
- Choose an enterprise that is suited to the size of your operation. You may not be able to raise cattle on 10 acres but goats will thrive on small acreage. Wheat, hay and alfalfa will take at least 150 to 600 acres while raspberries can produce the same if not more income from 1 acre. You can make a surprisingly livable income from just a few acres.
- Thoroughly investigate your potential enterprise. Talk to the local extension agent and get his advise on what products are suited for the farm or ranch you have either already purchased or intend to buy. Talk to breeders and farmers in the area regarding the pros and cons of local plants and animals. Research, research, research. After selecting a few enterprises that interest you and are suited for your location an in depth study program would be advisable. The Internet offers a vast amount of information on all agricultural products. Some of the more popular books available on these subjects include a line of books from Storey Communications including titles such as: Raising Poultry The Modern Way, Raising Goats the Modern Way and so forth. For those who want to farm Successful Small-Scale Farming An Organic Approach by Karl Schwenke is an invaluable guide to selection of equipment and raising, harvesting and marketing a variety of crops.
- Take a small farm approach to your business. You are not competing with the larger farms in your area. Do not produce the same products or more products than you can handle. Or buy equipment that is to large and inputs that are not appropriate for small-scale production. Since you will probably be working a full or part time off-farm job. Reasonably estimate time and financial resources. To try to imitate the larger surrounding farms could be disastrous. These farms have the past history, financial resources, labor force; equipment and technology appropriate for their size.

Labor

After choosing a product you may want to take an inventory of labor assets. Examine your resources and capabilities. If you are fully employed you may want to stick with less time intensive programs. Farming tends to be busy in the spring and summer. Ranching can be busy year round particularly in the fall and spring. Will your current employer give you the time off for harvest or birthing? Will he let you make up the time in the slack periods farming and ranching go through? Is it possible for you to farm full time while your spouse works to provide off-farm income? Do you have family help? What are the ages, physical and mental capabilities of everyone in the family and are their tasks suited for them. Does the family have the ability and discipline to work hard towards a common goal or vision?

In the beginning farm and ranch work may seem overwhelming. Until you get the farm operating in an efficient manner, the first year or so may leave you and your labor resources stretched thin. Once fencing and pens are up, equipment and machinery are purchased, all the quirks worked out and crops and livestock are on a schedule the farm will seem more manageable. The work may seem taxing but I know individuals that pay $100 a month to go to the gym to exert the same physical energy for nothing!

Location

Farm location will be the next consideration. Economic reality set's in. Unless you have come into a great deal of money you have 2 options. A larger farm far out or smaller farm close in. The further from an urban or suburban setting the less you will pay per acre and the more land you can afford to buy. If your employment requires you to live close to the city per acre the price will rise.

The type of farm product and means of marketing will also be a determining factor in the location of your farm or ranch. If you raise perishable crops for the fresh market you will need to live close to a population center. Hay and animal feed on the other hand can be stored, sometimes outdoors, for long periods of time and then shipped. Raising livestock can pretty much take place anywhere because they are generally transported live to market. Enterprises such as Pick-Your-Own will require a population radius around your farm. The farm then needs to be easy to locate and have good access.

Land

The land on which you farm or ranch needs to be assessed. The size of property to fit your agricultural endeavor. Do not go into farming with any preconceived notions about how many acres a farmer should have but what your particular circumstance is. If you feel you need lots of land to raise cattle, hay or wheat you maybe surprised to learn that many of the new technologies allow these enterprises to run efficiently and profitably on fewer and fewer acres.

For instance some cattle breeds have decreased in size. Dexter cattle have been genetically engineered to be smaller less consuming animals while giving comparable meat portions. Advancements in grazing management are allowing more heads per acre. With all the new small farm technologies and products buying large land makes little sense. Contemplate how much land you really have time for? Fences need to be maintained and larger equipment is needed for larger acreage. Could you buy less land and spend the savings on better equipment or more livestock? "The mistaken ambition of owning twice as much land as own can thoroughly manure or profitably cultivate, is the great agricultural sin of this country" from Ten Acres Enough: The Small Farm Dream Is Possible.

Soil and pasture conditions of the land are very important. You can spend a great deal of time and money trying to correct problems. Worn out or overgrazed land can take up to 5 years to regenerate.

Your best bet is to have an assessment done by a local farming authority. The ideal soil for growing crops is sandy loam. It should be rich in organic matter with a neutral pH and at least 6 inches deep. The land should gently slope to provide for run off. The topography should be such that it is protected against strong winds that blow seeds. If you will be Organic Farming soil that has been treated with the over-utilization of chemicals and fertilizers may interfere with your certification.

Water

Water sustains life. It is a necessity basic to crop and livestock production and should be closely scrutinized as part of the farms inventory.

Questions to ask about the water supply:

1. Is the quality good? Is the source supplying a good, constant supply of non-polluted water?
2. Is the water surface or subterranean? What are the irrigation rights? Can you irrigate and how much? How large is the land surface to be watered?
3. How many wells and how deep are they? How old are the wells and what is their condition. How many head can each well water?

Facilities

You know you have a true farmer's spirit when you are more concerned with the condition of the outbuildings and barns than the house. To build or renovate the facilities to meet the needs of your enterprise will require additional time and money. Many older farms currently have existing facilities already in place to accommodate hogs, cattle and poultry. Or buildings available for machinery and grain storage. Even though the conditions of the buildings may be deteriorating some shoring up and a little paint is much more affordable to the new farm budget that the cost of new construction.

Another consideration is the numerous small to medium size farms have more than one house on the property. Many families have established them selves in farming by splitting the cost of the entire farm with relatives.

Machinery, Equipment and Inputs

Low overhead and good management of resources can sustain the farm indefinitely. It is said the most powerful muscle in the body is the brain. Careful thought and planning is the strongest asset to the farm. Carefully contemplating purchases and debt load leads to better risk management in agricultural enterprises. To sustain a small farm the farmer either must increase profits or utilize less costly technology and inputs or preferably both to thrive.

Focus on keeping expenses low buy implementing the following example practices:

- Don't buy feed at the local retail feed store. Purina and other name brands make a lot of money distributing and mixing feed in those expensive, printed bags. Buy feed by the ton at wholesale from granaries. Price for cracked corn at a trendy feed store can be as high as $6.75 for a 50# bag. Cracked corn at the grainery can go for as low as $3.00 per 50#. Inform you grainery manager that if he cuts you a deal you will do all your feed business with him. Deals in hay can also be found by dealing direct with the farmer. Picking the bails up in the field can cut feed bills by up to 50%.
- Study up on veterinary medicine. Most vets will sell you the necessary drugs to treat animals without coming in for a visit.
- Local hardware stores and lumberyards often discount odd sized building materials. I once bought a whole truckload of "left over" pieces of plywood from a large chain for $5.00. They made nice pig huts. Although pieces of plywood may be of different thickness a coat of paint makes everything seem to even out. Besides the pigs didn't mind. The weather will pound, soak and blow on the outside of your facilities. Animals will kick, chew and push their way through the inside. It would be foolishness to waste profits on expensive esthetics.
- Auctions and private ads are a good source of used equipment, machinery and building materials.
- Utilize sweat equity when ever possible. Farmers will offer free fencing in return for taking it down. Same goes for barns and sheds. Farmers have bartered

and traded services, outputs and assets for centuries. Acquiring mechanical and construction skills would be prudent and a great asset to your farm enterprise.

- Use suitable technology. Before purchasing equipment and vehicles contemplate how much use you will be getting out of it. When I bought my 1945 Oliver Tractor, equipped with a front-end loader, my family thought I was nuts. I promptly explained tractors are not like cars. They are not rated by age or miles but by hours. If the tractor was originally on a small farm it was mostly likely only used for 3-4 months out of each year. I get about 10 hours of use each month. I use the tractor to move heavy loads, stretch fence, clean up pens and shovel snow. I paid pennies on the dollar for what a new model would have cost me and I get the job done in the same manner.
- Debt can be utilized as a farm tool. Used in the right way can lead to a productive outcome. Used in the wrong way can lead to destruction of your farming enterprise. If you are applying for a first time farmer loan from the state or feds consider asking for a reasonable minimum amount. Therefore the payments will not be such a strain on the farms economy in the first few years. You can always reapply for more and besides you are more likely to get your loan funded asking for smaller amounts.

Price, Profit and the Market

If marketing ostriches and EMU's does not ring your bell you may be interested in farming or ranching in a traditional farm product. Numerous traditional agricultural enterprises are adaptable to small acreage tracts. Both livestock and crops can be compatible to acreage's ranging from 6 to 160 acres. The advantage to traditional products is that the means for distribution have been well established and profits are often more predictable. When a farmer or rancher says he has sent his crops or livestock off to market it is usually implies sending his product to the local auction, cooperative or processor.

The following are typical traditional agricultural products that can be supported on a small farm...

Beef Cattle

Surprisingly their are many small acreage beef cattle enterprises. In fact the national beef cattle herd size is 25 head. With new "intensive" and "rotational" grazing techniques and methods having a herd of 25 or more on a small acreage is do able.

In general there are two types of beef cattle enterprises. "Yearling" and "Cow/Calf" operations. An old cattle rancher in my area told me the goal in any beef cattle operation is to make at least $100 a head.

Cow/Calf operations are the more profitable of the two types of beef cattle enterprises. It is also the most labor intensive. These operations breed their cows in late June and "calve out" the following spring. On average it takes about $200 to raise a calf to the yearling stage, realizing a $175.00 a head profit.

Yearling operations are intended to be a quick turn over of investment. Yearlings, or year old calves, are typically purchased in the spring at a weight of about 500 pounds, at about 75 cents per pound or $375.00. They are then turned out on pasture for about 100 days until they reach a weight of about 800 pounds. The calves are sold about 90-150 days after purchase depending on the cattle market for around 60 cents a pound. Therefore you could realize a profit of $105 a head. There is less risk in buying year old

calves but then there is less profit.
*Cattle prices are an example as the cattle market can fluctuate widely.

If you really want to cattle ranch but do not have the $500 to $800 per head to invest their is a cheaper way to operate a beef cattle enterprise.

Dairy operations are continuously calving out to increase their herds and increase their milk output. Only the female cows are needed to give milk. The bull calves are not necessary. Only one bull is needed to breed their many cows. Therefore most dairies sell their 3-day-old calves. The going rate is about $10-$20 a head for a Holstein bull calf.

Be forewarned.... A Holstein steer operation is very time consuming. You must bottle feed the bull calves for up to 6 weeks, twice a day! Milk Supplement and bottles can be purchased at the feed store relatively inexpensively. Once they are weaned the calves should be banded (castrated and made into steers, relatively cheap and easy with a store bought bander) and placed on pasture. A Holstein Steer is generally slaughtered at a year. Their muscle mass stops growing at a year with only the bone structure growing from then on. At market Holstein Steers bring about 55 cents a pound and are normally sold at around 700 pounds. A $385.00 profit from about a $55.00 investment.

After beef cattle are taken off pasture and sold at about 800-900 pounds they are sold then taken to the feedlot to be fattened, usually on grain, out to a weight of about 1075# for steer and 975 pounds for heifers. They are then slaughtered.

Sheep

Sheep are of relevantly low investment and are inexpensive to maintain. There are more markets for sheep products then there are for cattle. Plain white wool, naturally dark colored wool, freezer lambs and slaughter lambs. Experts say it takes about 2 hours of work per year to maintain one ewe and her offspring on farm pasture.

Sheep can be fed out on forage alone. Requiring little outlay for feed. Sheep unlike cattle that need pasture can forage. Abandoned cropland, woods and orchards can pasture sheep. As long as they are kept away from predators such as coyotes or dogs.

The initial start up costs of breeding ewes can go anywhere from $100-$150 each and rams about $200 each .....A typical budget for a flock of sheep is $51 per year in costs. Out of 100 ewes and average of 130 lambs are born. Freezer lambs are generally sold around 110# at about .99 cents per pound. Plus earnings from the wool the 8 to 12 pounds of fleece produced at sold for about $1.70 per pound.

Swine

Pigs will out produce just about any other farm animal. They can be a very profitable part time operation. Many a country kid has put themselves through college with pigs.

There are many advantages to pigs over other farm animals.

- They are early breeders typically at 5-6 months of age.
- Have 2 litters a year with the average gestation being 3 months, 3 weeks and 3 days.
- Average 10 to 24 piglets at each birth instead of the typical single to triple births found in other farm animals.
- Can go to market in only 5 months - a quicker turn around on investment.

- Cheap to get into. Piglets can be purchased anywhere from $20 to $60 each.
- Easy to care for. Pigs are hardy animals that adjust well to most environments and require little space. You can place many a pig on small acreage.

Typical budget for 10 pigs:

One pig eats about 500# of feed up to 6 months. They can be fattened out on just about any type of cheap grain. Cracked Corn plus additional supplements can be purchased for about 8 cents per pound.
500 pounds x 8 cents per pound = $40
$40 x 10 pigs = 400.00
Additional supplies include 5.00 for wormer, utilities misc. x 10 = $30

Total investment per pig is $43.00.

The average price paid for a pig at market is about 50 cents.

6 months to a typical market weight of 240# x .50 = $120.00 per pig
Ten pigs would bring $1200

Less investment to feed pig and additional costs=43.00

$77.00 profit on each pig

Investment of $430
Yield $1200.00
Return of $770.00 for 5 months.

*Pig budget was based on the use existing facilities and equipment. Minimal, part time labor was not figured in since the time it takes to care for pigs can be diverted from non-productive recreational hobbies such as watching the game or window-shopping. Other methods of feeding pigs can lead to greater profits. Jails, restaurants, bakeries and food establishments can be a great source of free pig food from leftovers and waste.

Or if you don't have the time or money to fatten pigs out you can start a "wiener" pig enterprise. Baby pigs generally stay on the mama for about 6 weeks. They are then weaned, thus called "wiener pigs". The only expense being a wormer, shots and feed for the mom. If you sold the piglets at 6 weeks at an average of $32.00, at an average of 10 piglets per mama, you would realize a profit of about $300.00. If you bred so that you had 2 sows delivering every month that would be $600.00 in income each month.

Pigs yield about 170# of meat and can save a family about $140.00 on their food bill. A typical after slaughter will produce the following:
40# of sausage
40# ham
20# bacon
70# ribs, roasts, chops and steaks

Vegetables and Fruits

Vegetable and Fruit producers have two distinct market outlets: fresh produce and processing. The fresh produce market generally produces the highest returns and is the most accessible to small growers. The processing market can provide a more stable income because it is less affected by over production and vacillating prices.

According to the USDA 1 acre of:

- Strawberries yields about 45,500 pounds with a value of $21,759
- Lettuce yields about 24,000 heads with a value of $5,940
- Oranges yields about 28,800 pounds with a value of $2,427
- Sweet Corn yields about 14,400 pounds with a value of $2,674.

Hay

There are two types of hay suited for small acreage production. Alfalfa and Grass. Both are used for animal feed.

Alfalfa can either be grown dry land or irrigated and can be cut up to three times during a growing season. Yields of 1.5 tons per acre are common. Once alfalfa is established the stands can last for up to 10 years. Returns on alfalfa range from $40 to $80 an acre.

Grass hays only have one cutting per year and are generally harvested in July to August. Grass hay offers less of a return than alfalfa at approximately $15 per ton per acre.

Most small farmers that produce hay sell to the local population of livestock and horse owners. Word of mouth or inexpensive local advertising to sell their left over inventory. Most small farmers generally hay farm to feed their own animals, realizing a greater profit in the sale of their livestock.

Trees; A Small Farm Investment Strategy:

Forget stocks and bonds as an investment strategy when you can use trees and do something good for the environment as well as your wallet.

Black walnut trees will bring as high as $5,000-$32,000 each when full grown at the end of 20 to 30 years. In the spring time your local Soil Conservation District will offer the seedlings for around $20.00 for 100 . Your $20 investment can turn into $500,000 in 20 years. I don't know any stock or bond that can out perform that!

New Paulownias Trees grow to 20 feet in the first year. Super-cloned to produce quality time in eight to 12 years. Can be a source of Mahogany or poplar like material, it takes a poplar 65 years to mature. Carolina Pacific International Inc. 111 Lindsey St., Lenox, GA 31637 800-706-5037 http://www.paulownia

To Learn Farming Is To Experience It

All the research, preparation and planning maybe key factors to the success of a farm but until it is a reality it will be largely unknown whether you will be able to adapt to it's vigorous, daring lifestyle. During my first year at the ranch the first winter storm to move in was a blizzard. After the storm cleared out, I was at the local hardware store purchasing some much-needed supplies. While at the checkstand I proceeded to tell the store clerk about all the hardships I had faced in the last 48 hours. It took me 3 hours on the tractor and about 5 gallons of gas just to get the beast out from around the barn to the driveway, which took me another 4 hours to clear out. I was caught unprepared running low on feed. After the storm, the drifts were so high that 80 goats simply strolled right over 6 foot cattle panels into the open field and needed to be rounded up and placed in other pens. After patiently listening to my belly aching the store clerk calmly turned to farmers in the other lane at proclaimed "It's her first year".

Chapter 7 Getting The $$$ To Buy Your Rural Home, Small Farm Or Mini-Ranch

Placing monetary and financial concerns aside for a moment, as they are important to the future of your family, one must reflect in the true profit and value of becoming a country dweller. A sense of safety, community, high levels of freedom, low stress, clean water and air is your best investment strategy.

You are planning your own Escape from New York, have selected your ideal rural estate in Refugee Acres, and are ready to make the move. Or you own an existing small farm or mini-ranch and want to start an agricultural enterprise. You will be in need of funds to purchase the real estate or obtain capitol for operating expenses such as equipment or livestock. There are many options to consider when seeking rural and agricultural financing.

Most rural and agricultural financial capital is generally provided from federally insured depository institutions, private firms or cooperatives, government sponsored programs and individuals.

It has not always been easy to get mortgage or business financing in rural areas. During 1992-1993 data from mortgage lenders, who reported subject to the Home Mortgage Disclosure Act, showed that the approval rate for home purchase loans were significantly lower in rural areas than in metro areas. Only 63% were approved compared to urban loans at 77%. These figures suggested serious credit access problems in rural areas. The government has been attempting to stimulate rural growth through it's many Rural Development Programs that address credit and qualifying issues. In consideration of credit demand one must recognize that the Federal Government is involved to some degree in all rural banking industries. Past intervention from the Federal Government in rural credit markets has addressed concerns over fairness, accessibility and has enhanced rural credit efficiency.

The best bet to find financing for rural or agricultural property is in the community, in which you will live or do business. Many non-rural lending institutions do not understand rural property and shy away from homes with acreage. This I can tell you from experience. When I was searching out a mortgage for my small ranch I was asked by several potential metro-based lenders why the home was not fueled by natural gas. I then had to explain to my mortgage broker who in turn explained to the various banks that it was not profitable for the Gas Company to run a gas line down a 10-mile dirt road for only 8 customers. Many of the appraisers that are sent by city based lenders are not familiar with estimating the value of barns, fencing and other farm/ranch amenities. Rural banking institutions have a better understanding of the value of farmland, real estate, equipment and livestock. In the event real estate or chattel must be repossessed or foreclosed the rural lending institutions have means and outlets to liquidate the assets therefore are more willing to finance these types of capital.

The requirements to obtain credit from these groups varies widely. Many factors are taken into consideration such as collateral, ability to pay, past credit history and eligibility.

The following institutions are a dependable source of financing for American Agriculture. They understand rural and agricultural investments and can provide the funds for country or rural mortgages for purchasing, building, refinancing and remodeling. Real Estate Loans for land, buildings, facilities and land improvements. Credit lines for operating expenses, livestock, feed. Leases and loans services for farm equipment, vehicles, computers and irrigation systems.

Commercial Banks Commercial banks are by far the most dominant and visible force in rural financial markets. 2/3 of lending institutions in rural area's are commercial banks and they control over 80% of the assets held by rural banking customers. Commercial Banks can provide long and short-term credit for a wide range of rural and agricultural uses including home mortgages, personal loans, agricultural operating loans and commercial loans. As the rural rebound movement has progressed, more people have moved to the country and the real estate holdings of rural commercial banks have increased. In 1990 home mortgages accounted for 27% of their total loan portfolio. At the end of 1995 that number rose to 31%. The commercial bank will most likely continue to dominate as a source of credit for rural borrowers.

Farm Credit System The Farm Credit System was created by Congress in 1916 to provide a dependable source of credit for the nations agricultural segment. Since it's inception Congress has expanded the System to better serve farmers, ranchers and the rural community in general. The Farm Credit System is a nationwide system of cooperatives that loan money to agriculture and rural America. The system provides more than 25% of the credit needs to US Agriculture. Currently the system provides 59 billion in loans to more than ½ million farmers, ranchers, cooperatives and rural utilities. Most of its loans are made directly to an individual farmer for real estate purchases and farm production. The Farm credit system, with its 8 banks and 228 associations, serves every region of the country and has grown to be a major source of agricultural and rural credit.

Savings And Loan Associations The Savings and Loan Crisis of the 1980's, compounded with mergers and acquisitions, have dramatically decreased the number of S & L's that serve rural and agricultural communities. Only 496 or 1/3 of the nations S & L's were headquartered in Rural America at the end of 1994. Although rural S & L's tend to be smaller they still average more assets than a typical commercial rural bank and average about $113 million in assets. This is due to rural S & L's placing more emphasis on home mortgages than commercial or business loans.

Credit Unions Credit Unions are not a major player in rural and agricultural communities. At the end of 1995 out of the 11,958 locations only 22% where located in a rural area. Credit Unions provide much of the same services as commercial banks do and their loan portfolios are comprised mainly of home mortgage and consumer loans. In the small percentage of rural areas they are located in they can be a source of real estate credit. Credit Unions very rarely make business or commercial loans.

Finance Companies Like Credit Unions, Finance Companies main lender focus is Real Estate and Consumer Loans. Finance Companies tend to target "high risk" applicants to which they charge a higher interest rate for funding the loan.

Individuals Relatives, friends and personal savings are in general the source for the majority of real estate down payments. In 1995 individuals held 23% of the outstanding farm real estate debt. Most of which were in the form of 1st and 2nd Trust Deeds. Paper taken back by the owner of the property can fill the gap of where conventional financing falls short. Individuals finance the majority of the raw land sales in this country. When purchasing raw land if the owner is un-willing or unable to carry paper or if the buyer does have cash in hand to make an out right purchase, he or she is in for an up hill battle. The great majority of the banks and lenders will no longer finance un-improved land, especially if your not planning to build right away. If you do approach a conventional lender be prepared to put about 40% - 50% down and pay a stiff interest rate.

Government

The basic function of government to provide a stable environment in which economic growth can occur. To stimulate rural development and to insure farmers and ranchers have a stable and reliable source of credit. Both state and federal governments have been initiating government-funded programs to provide assistance and fill in the gaps that commercial credit leaves in the system.

Government Rural Housing Programs Guaranteed Rural Housing and Rural Direct Loan Leveraging Programs. These programs provide borrowers in rural areas easier access to affordable housing finance options. Fannie Mae has teamed up in conjunction with the USDA Rural Hosing Service to provide low and moderate-income rural residents with better access to affordable credit and decent housing. 1.7 billion in mortgages for the fiscal year 1997 for middle income rural homebuyers. The benefits to the borrower are no down payment and up to 100% financing. In addition to a below market interest rates on a 30 year fixed loan to purchase a home or for new construction of a home.

To qualify you must buy a home in a small town or community of 10,000 or less. Qualification includes earning requirements; a good credit history and the PITI cannot exceed 29% of gross monthly income. These programs are similar to VA home loan guarantee programs. For more information on this program you can contact the local US Department of Rural Development/USDA - Rural Housing and Community Development Office located in Chapter 8.

Farmer Mac Loans The Federal Agricultural Mortgage Corporation was created in 1987 with the passage of the Farm Credit Act. Created by Congress, Farmer Mac attracts new financing capital for agricultural real estate and rural mortgages. It provides farmers with a secure and reliable source of credit and has the ability to access a 1.5 billion-dollar direct line of credit from the US Treasury. Interest rates are competitive with the commercial market and loans can be amortized up to 25 years.

Beginning Farmer/Rancher Programs "America's prosperity and vitality depend a great deal on our nation's continued leadership in agriculture. To maintain this leadership into the 21st century, we need to make sure we have new farmers moving into the field to replace those who will be retiring. To make this happen, we must find new ways to encourage and assist young farmers to overcome barriers that prevent them from starting their own operation. This program is a step toward this goal. I urge young farmers just starting out to look into these programs. It may be just the thing many of them need to start their own operations on the right track". Dan Glickman, Secretary of Agriculture expressing his view on the Federal Beginning Farmer/Rancher Program administered by the Farm Service Agency of the Department Of Agriculture.

This sediment has also been held by individual states that have followed suit with their own programs to promote the vitality of agriculture in their own backyards. Many states have developed special agencies or departments known as Agricultural Development Authorities to administer these programs. These programs are intended to stimulate interest in farm and ranch ownership.

Listed in chapter 8 are the Federal Agencies that currently offer special loan and financing programs for Beginning Farmers and Ranchers. State programs are usually administered by a State Appointed Agricultural Authority or by the individual states Department or Office of Agriculture located in Chapter 10.

Many states have programs under development. Periodic checking with the Department of Agriculture of that state or The National Council of State Agricultural Finance Programs (NCOSAFP) can tell you if a program may be in the works. The NCOSAFP can be contacted at: NCOSAFP Steering Committee, Iowa Agricultural Development Authority, Wallace State Office Building, Des Moines, IA 50319 (515) 281-6444, Fax (515) 281-6236.

State Agricultural Programs

The purpose of beginning farmer/rancher loan programs is to assist beginning and first-time farmer/ranchers with the purchase of land, real estate, equipment and breeding livestock.

Here is how state operated Beginning Farmer/Rancher Programs works:
A state develops a beginning farmer/rancher bonding program and can issue bonds to private investors or the party that will loan the money. These bonds earn interest and that interest is federal and/or state tax free to the investor/lender. In other words the private investor/lender pays no tax on the interest earned on the loan. Many states use this process to finance industrial and housing projects. The investor on the loan earns, generally, more interest than he would in a savings account, but the interest is tax-free. The same incentive is offered under the beginning farmer/rancher programs. The qualified beginning farmer/rancher and the private lender coordinate the loan with the state. It is a win situation for all three parties. The beginning farmer/rancher receives interest rates on the loan an average of about 2 percent below commercial rates. The investor or lender earns above average interest on his investment, tax-free. The state, to which most are short of expendable cash, puts out no actual moneys and stimulates it's agricultural economy.

In addition to the standard agricultural loans, many states offer different types of special agricultural financing or loan guarantee programs that target the first time farmer/rancher. Some states offer special loan programs for agricultural processing or various types of agricultural enterprises other than farming and ranching.

Chapter 10 lists individual states, which participate in agricultural development programs. Qualification rules and the approval process for these programs vary from state to state. In general the following are rules for eligibility in these programs. Contact the individual state departments for the specifics.

To qualify as a beginning farmer/rancher you must:
1. Have no past history of ownership of a substantial amount of farmland.
2. May not borrow more than $250,000.
3. Your net worth must be no greater than $300,000.
4. May not purchase the property from immediate family or closely related persons.

Federal Agricultural Programs

The USDA supports the profitability and productivity of farming and ranching through their Farm Service Agency. For over 60 years the Farm Service Agency's goal has been to preserve and promote American Agriculture. United States agriculture generates One Trillion Dollars towards the economy each year and is 15% of the gross domestic product of this country. American agriculture provides 22.8 million jobs for Americans. It is important for the nation to keep agriculture a viable enterprise now and in the future.

The Farm Service Agency The Farm Service Agency helps the American Farmer/ Rancher with credit, commodity, conservation, crop insurance, export and risk management programs. The programs administered by the Farm Service Agency help keep farmers and ranchers in business. These programs improve economic stability of agriculture and help farmers adjust to meet new demands. The FSA aids farmers and ranchers in stabilizing their income, conserving water and land resources, recovering from the effects of natural disasters, and provides loans or credit for beginning, new and disadvantaged farmers and ranchers.

Farm Service Agency Loans The Farm Service Agency offers direct and guaranteed farm ownership and operating loans to those farmers unable to obtain credit from local lenders because of insufficient credit background or networth. Also, farmers with limited resources to start and maintain a successful farming, farmers and ranchers that have been hurt by natural disasters and farmers and ranchers who are socially disadvantaged.

These loans include: Farm Ownership Loans, Farm Ownership Downpayment Loans, Farm Operating Loans and Loans Targeted to Beginning Farmer and Ranchers or the Socially Disadvantaged. The interest rate on FSA loans are generally from 2% to 6% below commercial interest rates. Repayment terms for direct loans are applied according to the borrowers ability to repay. Loan terms can be as short as 1 years and as long as 7 years with some loans being extended out to 15 years.

Beginning Farmers and Ranchers

Each a year a portion of the farm ownership and farm operating loan funds are set aside by Congress for Beginning Farmers and Ranchers. A qualified Beginning Farmer or Rancher is an applicant who:

- Has not operated a farm for the preceding 10 years
- Will actively participate in the day to day management and operation of the farm or ranch.
- Must participate in loan assessment and borrower training.
- Must not purchase or own a farm or ranch more than 25% of the average farm or ranch acreage in the county where the property is located.
- Must posses the skills necessary to enter and operate a viable farm or ranch operation. Your background does not need to be in agriculture to qualify.
- Meet the FSA ownership loan eligibility requirement.
- If the applicant is an entity, all members must be related by blood or marriage, and all stockholders in a corporation must be eligible beginning farmers or ranchers.

Funds for the Socially Disadvantaged are also reserved each year. According to the USDA A Socially Disadvantaged Farmer or Rancher is "one of a group whose members have been subjected to racial, ethnic, or gender prejudice because of their identity as a member of the group without regard to their individual program, socially disadvantaged groups are women, African Americans, American Indians and Alaskan Natives, Hispanics, Asians and Pacific Islanders.

The local Farm Service Agency offers loans to assist Beginning Farmers and Ranchers and those that are Socially Disadvantaged enter agriculture and at the same time provide additional avenues for older farmers to transfer their land to future generations.

It is possible for an applicant to be both a Beginning Farmer/Rancher and Socially Disadvantaged.

Beginning Farmer/Rancher Down Payment Loan Program: Under the Downpayment Program, the USDA will finance 30% of the land's purchase price over 10 years at a fixed rate of 4%. The loan applicant must make a cash downpayment of 10% of the purchase price with the remaining 60% financing to come from a commercial lender or private party. The USDA can now provide eligible lenders with a 95% guarantee on the remaining 60% of the purchase price. State agricultural programs can also be used to finance the remaining portion of the loan. In several states blending the financing is being successfully used to an attractive low interest rate to First Time Farmers and Ranchers combing the FSA loan with the states loan that fall well below commercial rates.

Beginning Farmer/Rancher Operating Loan Program: The USDA Farm Service Agency also offers an operating loan program for Beginning Farmers and Ranchers. Eligible applicants may obtain a direct loan from the FSA for up to a maximum of $200,000 and up to $400,000 for a guaranteed loan. Most loan terms are up to 7 years. Loans can be used for normal operating expense, family living expenses, machinery and equipment, real estate repairs, improvements and refinancing of existing debt. Like the Down Payment Loan Program applicants must demonstrate the ability to caring out a detailed 5 year operating plan. Applicants must also own the operation or have a lease commitment and meet other eligibility criteria.

Loans For Socially Disadvantaged Persons The purpose of these loans are to assist socially disadvantaged persons with direct and guaranteed loans for farming or ranching by removing obstacles that prevent the full participation of these persons in FSA farm operating and ownership loans. Farm ownership loan funds can be used for the purchase of and to enlarge a farm or ranch, purchase rights of way or easements and to erect or improve buildings Operating loans can be used to purchase livestock, equipment, feed, seed, fuel, chemicals, fertilizer, insurance and labor. Interest rates on these loans are set according to the Governments cost of borrowing the money and usually run from 1 to 7 years. Ownership loans can be set as long as 40 years.

Farm Ownership Loans Guaranteed farm ownership loans to operators of family sized farms. Maximum available for direct loans is $200,000 and for loan guarantees is $300,000. The county FSA committee determines the eligibility of the applicant. Interest rates are generally lower than commercial rates.

Farm Operating Loans Farm Operating Loan proceeds can be used to pay for item need for farm operations such as livestock and equipment. Terms on these loans generally range from 1 to 7 years. Funds are limited to $200,000 but can guarantee loans for up to $400,000 with interest rates below current commercial rates.

Supervised Help and Loan Servicing After placing applicants on their farms or ranches the FSA offers the means to keep them there and successfully in business. Two means the FSA uses to accomplish this is Supervision through out the loan process and servicing of loans that may become delinquent due to circumstances beyond the applicants control.

All farm loans through the FSA are supervised. This means that the FSA works with each borrower to identify specific strengths and weaknesses in farm/ranch production and management, then works with the borrower on alternatives and other options to addresses the weaknesses and achieve success. Effective supervised credit is the difference between success and failure for many farming and ranching families. The FSA provides the help to farmers and ranchers to use their resources most efficiently and run their operations profitably.

Most farm or ranch loans are set up for a lump sum yearly payment. If these payments can not be met the Farm Service Agency has many options. They can re-amortize, restructure or defer the loan. Lower the interest rate. And can accept environmentally sensitive land in exchange for a writedown of the loan.

How to apply

Applications packages may be requested at any local Farm Service Agency listed in Chapter 8 or are available in downloadable form at http://www.fsa.usda.gov/dam/forms/farmlnfrm.htm. In most cases you will be required to submit the following information and documentation to the Farm Service Agency in order to complete the application process:

- A completed "Application For FSA Services"
- If your application is for a cooperative, corporation, partnership or join operation you must submit a complete list of members, stockholders, partners or join operators, showing their addresses and percentage of ownership or interest held by each member. A current financial statement of each of the members, a current financial statement from the entity itself and any agreements or articles of incorporation.
- A business plan that includes the following: a brief narrative as to farm training or experience, development plan, details of the proposed operation and the proposed size of the operation, details describing the projected production, income and expenses, loan repayment plan and proposed financials.
- Personal financial records for the past 5 years.
- A copy of your lease, contract, agreement with any landlord and the legal description of the property.
- Deed showing a legal description of any property owned.
- Written evidence from your present lender or other local lenders documenting your inability to obtain credit.
- Credit References.
- A credit report fee.
- Complete "Farm and Home Plan"
- Verification of Employment
- Highly Erodiable Land and Wetland Conservation Certification.
- Request For Environmental Information.
- Application Certification - Federal Collection Policies For Consumer Or Commercial Debt.
- Request For Statement of Debts and Collateral.
- Applicant Reference Letter List
- Statement Required by the Privacy Act.

Your Farm Service Agency Supervisor can help though out the whole application process with obtaining the necessary paperwork to complete the process.

Security Required The Farm Service Agency will require a lien on the applicant's assets. This may include all or part of any Chattel: crops, livestock and equipment. And/or real estate.

Farm Service Agency Loan Approval Process Once an applicant has applied and submitted all the necessary information to their local FSA office the application is reviewed by the FSA County Committee who determines the eligibility of the applicant under the law. An applicant who is found eligible for assistance does not automatically get the loan. It is then determined whether the applicant meets all the requirements for loan approval. Then to approve the loan the FSA County Supervisor works with the applicant and must decide whether the applicant will generate enough income to meet all expenses of the farm or ranch plus repay the loan and any other debts. Local FSA Offices generally take up to 60 days once the application is submitted to process loan applications. Once the loan is approved the applicant is generally notified by mail and a closing date is set.

Applying For R_ _l Estate Credit

The Loan Process Before shopping for a rural property is it best to know what you will be able to afford. Falling in love with your dream ranch and then finding out you are unable to qualify for the payment can be a heartbreaking experience. To pre-qualify or pre-apply is a process in which it is determined what amount of mortgage or price of a home you can afford.

The pre-qualifying procedure normally consists of:

- Credit History - A Credit History is run upon signing the initial loan approval form. If problems exist on the credit report that negatively reflect on your loan you will be notified and possibly be asked to clear the up the problems with the reporting agency.
- Uniform Residential Loan Application - In order to complete this form the borrower will need banking information including account numbers and balances, asset information including homes, cars, insurance policies and any other significant asset and names, addresses and accounts numbers of creditors.
- Additional forms to complete for pre-qualifying: Application Acknowledgment, Appraisal Disclosure, Fair Lending Notice, Authorization For Credit Release, Work Order For Title Inspection, Statement Of Owner Occupancy and Servicing Disclosure Statement.

Loan Programs Most lending institutions offer several loan formats.

- Fixed Rate Interest rate remains constant or the same during the life of the loan.
- Adjustable Rate At predetermined intervals, the interest rate is adjusted up or down to reflect changes in market rates.
- Variable Rate The interest rate is adjusted to accommodate changes in cost of funds, competitive factors or other considerations.

Factors such as credit history, ability to pay, length of employment and collateral will be the determining factor for the amount available to finance.

The Loan Process will proceed as follows:

- Pre-Qualification - Examine credit history, determine debt ratios, define loan program to fit borrowers need, determine ability to pay.
- Formal Application - Submit Uniform Residential Loan Application and all additional forms.
- Assemble Information For Processing - Last 2-5 years of tax returns from applicant and co-applicant, 3 months of Bank Statements, Pay Stubs, documentation showing where down payment and closing costs are coming from, miscellaneous paperwork (Divorce decree, receipts for child support, etc.).

- Underwriting - The final determination as to whether the lender feels the borrower has the ability to pay the loan back based on all the above-submitted information.
- Closing - A formal signing of all documents, settlement statement, the note and deed of trust. All moneys to and from all parties plus settlement fees are transferred.
- Funding - Directly following the closing the new deed is recorded and the loan is funded.

Applying For Business Credit

In the pursuit of funds to either start or expand any agricultural project a business plan is almost always required. Most lending institutions such as the local bank, government agencies such as the Farm Service Agency or State Agricultural Divisions will require some sort of written plan detailing where you are, where your going and the details of how you intend to get there.

Solid planning sets the stage for a successful funding outcome and helps any business owner focus on direction and goals. It is a road map, telling the story of your business, where you are starting from and where you will end up. Your business plan tells the reader that you are serious about your project, that you have put much thought into the operations, marketing and financial planning both now and in the future.

To obtain funding you must prove to the lender you have clearly thought out and planned how and where you will obtain the means to pay them back and that you are a good risk. A clear concise Business Plan can be the make it or break it point in obtaining funding. If the lender is impressed with the knowledge you have in your agricultural field and effort you have put into your Business Plan he will assume that you will put as much effort into executing that Business Plan into reality for your business. Make sure your message is clear so the lender cannot fail to understand why they should finance your project.

Anyone sitting down to write a Business Plan can get overwhelmed by the all the detail and where to place it. You're best bet is to start with an outline then work into fitting the pieces together.

Your outline should have only one main idea for each paragraph and begin each paragraph with a sentence that states the main idea. Organize each paragraph from general to specific, using key points in lists to clarify and simplify your ideas to keep the lenders attention.

Whether you type your plan on a computer or write it on a pad of paper, it will require a lot of polishing and exhaustive review. The end result needs to be very persuasive and convincing.

Your business plan must concentrate on 3 specific areas.

1. You're current position.
2. Details regarding the execution of your plan.
3. What you will require to successfully execute your plan.

#1 Your Current Position

Your current position should include all your assets both financial and personal. A complete detailed analysis of your current financial situation including credit history, a balance sheet, cash flow analysis, profit and loss, financial projections and a sales forecast. This section should also detail what type of business you have, a corporation, partnership or sole proprietorship. If you have a corporation list the Board of Directors and Shareholders.

Outline all your skills related to running the purposed operation, just as if you where completing a resume or job application. Anything pertinent to the type of operation you are proposing. List any accounting, marketing or management experience or knowledge you have acquired in the past that proves.that you are competent to achieve the successful outcome of the proposed plan. Any educational knowledge will also be helpful. List all college and any courses you have taken that relate to your business.

Next outline a detailed accounting of your operation: how many animals you have, the types and sizes of existing facilities. What types of machinery you have now. The crops your are growing and harvesting currently. Any and all current assets you have that are held personally and in the business.

#2 Executing Your Plan

Determine in what direction do you want to take your business. What are your overall financial and growth goals. What is the timetable to achieve these goals? List as many details as possible. It is helpful to list all your goals first then next to them list all the ways you plan to reach those goals. If you don't already have an existing operation, write your plan as if the operation actually exists.

You will need to show what type of condition your facilities and equipment are in and the repairs that will be necessary to put them in a working state and what you purpose to do with them. If you are starting from scratch you will need to investigate properties available in your area that fit your needs and what they are selling for. The same goes for equipment. List your hours of operation, especially if you work a full time job in the interim, the lender may want to know how you intend do both. If you plan to start very large scale you may need to employ some help. If you plan to employ help, what kind of salary and benefits you will provide.

You will also want to detail your product or product line. Think hard about your end product. Keep an open mind to your product and where it could lead. Several farmers and ranchers are not only selling wholesale but are also marketing value-added products. For example if you were to raise strawberries, what is your market? Wholesale, direct from a fruitstand, a pick your own operation, jams, syrup, candy?

The 3 keys to understanding your market are: Your niche, competitors and customers. Where you will sell your end product and for what price. Will you sell your product directly to retailers, wholesalers or to the public or possibly to all three. Where you will sell your product and to whom will be greatly determined by what your output is. If you have a large ranching operation you will most likely sell wholesale to a livestock yard.

Give an example of another business in your related field. List any and all competition and what they are doing in their business to get their end product to market. How will you be different than the competition? Discuss advantages and disadvantages of competing products and where your product will be different. Do some market research of your own. Call around and ask prospective clients if you could supply them with a product at an attractive price would they buy from you?

Develop the link between you and your customer. In order to do this you must properly define your market niche. Understand your critical success factors. Develop your sales forecasts and prepare an action plan to penetrate your market.

Some thought must be given to advertising. You must determine how you will reach your customers. Most business cannot just depend on word of mouth. Letterhead and business cards can be a source of advertising so some effort on their creation must be made. Attending livestock shows and joining different associations can be very helpful in agricultural fields. Belonging to a group such as The American Boer Goat Association can not only promote sales of your breading stock but it's members are usually very helpful and can offer advise as to where to market and sell your products.

A market analysis should be provided to the lender. It should be a descriptive of the total market size and growth rate. A review of previous year's sales and growth rates along with projected growth rate for your industry will help the lender understand that past and present growth in your particular industry.

 # 3 Funds Required

Next you will need to determine the purpose and use of the funds you are asking for. Be specific as to how much money you will need and where the money will go. Determine your start up and initial operating costs. Your start up costs should include: beginning inventory, equipment, facilities, repair, maintenance, advertising and promotion, insurance, salary, utilities and taxes. Project your monthly expenses for each item.

Briefly describe how the money will be used. You will probably use the funds for: product development, equipment, marketing, working capital, inventory and the purchase or lease of property and facilities. It is easy to get carried away here. Remember you will obligated to pay this money back. If this is your first attempt at agriculture.....be frugal! And remember the less you ask for the better chances you will have of getting the funds.

The funds required should include the amount of funding, the amount and type of security offered to the lender to secure the loan, the uses of the capital and how the capital will be paid back. Based on your cash flow you will need to prove what portion of the proceeds will be used to pay back the loan and what proceeds will be used for daily operations. The lender will want specifics on collateral used to secure the loan. Describe the securities offered. The specific type, values and condition.

One thing to keep in mind is that once the lender gives you the money it is still their money until it is paid back, with interest. They will be very concerned about what ratio is going to be spent on collateral and what portion will go to expenses. Most lenders will either secure the loan with real estate or chattel (livestock, vehicles, equipment or crops). This could include livestock or equipment that is not yet purchased. The lender will want to know the specifics as to how much per head and what kind of quality livestock or equipment you will be purchasing. In the event they must take it back and sell it.

Example Business Plan

The following is an example of an agricultural business plan that was submitted and successfully funded by the Farm Service Agency. Research and financial calculations will be necessary to complete a business plan tailored to fit the needs of your enterprise.

Business Plan

This business plan contains proprietary information that is not to be shared, copied, disclosed, or otherwise compromised without the consent of W A Ranch.

1.0 Executive Summary

1.1 DESCRIPTION

The following is a business descriptive for a new agricultural venture called WA Ranch. The company was formed with the objective of becoming a viable and profitable small scale ranching establishment by producing diversified livestock products that can be sustained on small acreage. To achieve it's objectives the company is seeking $20,000 in funds to start, maintain and increase the operation thus increasing profit and providing for the success of the enterprise over a 5 year period. This enterprise will be managed and organized in a creative and innovative fashion to generate very high levels of efficiency in the production and sales of livestock products. Currently the economic environment is favorable to the production of the proposed agricultural livestock products. By utilizing current facilities, knowledge of key personal and strategies outlined in this Business Plan, WA Ranch, will become and maintain a position as a successful small scale agricultural endeavor.

1.1.1 Nature of Products

The livestock products that will be produced will be such that a small scale ranch can manage and sustain in a profitable manner. Goat Meat, Cashmere Hair and Pheasant will be the main product line along with other products that maybe related to such products.

1.1.2 Unique Features of Products

All products can be effectively and profitably produced on a small amount of space with the proper management. Goat and Pheasant Meat are increasing in popularity as the health conscious public becomes aware of alternative forms of meat intake. The large influx of ethnic goat eating population is creating a great demand. Both meats have sustained and increased their demand and pricing in recent years. Cashmere Hair has always been treasured worldwide and demand has continued to rise. All items are currently being imported to the United States but with associations and groups continuing to organize their efforts to develop new in roads to get these products to market prices continue to stabilize and increase.

1.1.3 Revenues/Growth Rate/Profit

The projected annual gross revenues and profits are as follows:

	Year 1	Year 2	Year 3	Year 4	Year 5
Revenues	$17,280	$25,841	$37,742	$49,643	$67,294
Growth Rate	50%	46%	32%	24%	26%
Profit	$12,771	$19,470	$28,952	$38,164	$53,127

1.2 STRATEGIC DIRECTION

WA Ranch is a startup company dedicated to profitable livestock production.

1.2.2 Long Range Direction:

The long range goal is to become a profitable small scale livestock producer with significant growth every year through the utilization of current information and technologies pertaining to alternative agriculture techniques and by utilizing the land, current facilities and knowledge of the owners to their fullest capacity.

1.3 MARKETING

1.3.1 Goat Meat Markets

Currently the meat goat industry has a product shortage. Much of the goat meat is being imported to the United States because most ranchers are unaware of this shortage and of its popularity in some population groups. Immigration to this country continues to grow at an unabated rate, thus causing additional shortages in goat meat. Consumption of goat meat has moved up substantially since the mid 1980's and has held more or less stable prices. Total goats slaughtered at federally inspected plants has more than doubled since 1980. Supply and demand are shifting upward indicating a growing industry.

1.3.2 Cashmere Goat Hair Market

The commercial market for cashmere has been around for centuries. Supply has always exceeded demand. Local Cashmere buyers and Cashmere Co-ops buy hair continuously at fair prices. See Distribution.

1.3.3 Pheasants

There are typically 4 major markets for pheasants. Gourmet food markets (upscale grocery stores, caterers and restaurants), private individuals that buy birds either for breeding or custom slaughter, individuals and groups that release the birds into the wild and hunting preserves.

1.3.4 Summary of Advertising & Pricing

1.3.4.1 Advertising - The following primary methods will be used:

The goat products will be sold at Texas livestock markets and the Cashmere Associations to maintain a constant cash flow. However during this time the company will make a tremendous effort to gain acceptance and popularity as a local producer of goat and pheasant meat. A special marketing and promotion effort will be made to place the meat locally in upscale restaurants, grocery stores, catering establishments and health food stores. Marketing efforts will be made to attract ethnic groups to buy goats directly from the ranch for custom slaughter. The same for hunters and hunting preserves. The management of the ranch is very familiar with self promotion, and has in the past received thousands of dollars worth of free advertising for other ventures. The same attention to self promotion will be directed towards this company.

1.3.5 Pricing

Current market prices.

1.3.1 Sales Projections

Year 1	Year 2	Year 3	Year 4	Year 5
$17,280	$25,841	$37,742	$49,643	$67,294

1.4 MANAGEMENT

1.4.1 Backgrounds:

Owner had over 15 years experience in marketing, management and administration. Has owned livestock for the past 4 years and comes from a farming/ranching background. Currently owns a small ranch with all the facilities to accommodate the proposed livestock.

1.4.2 Responsibilities

Administration Management (Book keeping and Financial Management)
Operations Management (Livestock Management)
Marketing (Promotion and Sales)

1.5 Financial Projections

	Year 1	Year 2	Year 3	Year 4	Year 5
Revenues	$17,280	$25,841	$37,742	$49,643	$67,294
Net Income	$12,771	$19,470	$28,952	$38,164	$53,127
Assets	$137,353	$161,236	$189,688	$221,083	$261,463
Liabilities	$13,289	$14,781	$17,580	$20,390	$23,211
Net Worth	$124,064	$146,455	$172,108	$200,693	$238,252

1.5.1 Capital Needed:

Purpose	Amount
Used Tractor/w Front End Loader	$4000
Fence	$5500
75 Breeder Does and 1 Buck	$6500
1 Year Feed & Expenses	$3000
200 Starter Chicks	$280
Repair/Renovate Barns	$720
Total	$20,000

1.6 FINANCIAL ARRANGEMENTS

1.6.1 Financial Terms:

Equity is held by sole owner. All income and financial

responsibilities are that of the sole owner.

2.0 BACKGROUND AND PURPOSE

2.1 History

2.1.1 Brief Synopsis:

Since this is a start up company, this section will be completed after the companies first year of business.

2.2.3 Current Conditions

WA Ranch as it exists now, already provides all the accommodations and facilities to properly maintain the proposed livestock. Currently such livestock as goats, horses and chickens are being sustained at the ranch. Buildings and fencing for the proposed livestock production are in the process of being renovated from their current condition. Office and book keeping equipment are also property of the ranch and are constantly being maintained and upgraded by the owner. Book keeping procedures and marketing literature are being created and set in place for the operation. All mortgage payments, utilities and taxes are currently being paid out of the owners income.

2.2 Concept

2.2.1 Key Success Factors

- Diversification of Product Line
- Low Overhead
- Marketing
- Innovation
- Growing Industry For Product Line
- Demand Historically Exceeds Supply For All Products

2.2.2 Unique Features Of Products

- All Livestock Can Be Easily Raised On Small Acreage
- Goats Are Dual Purpose - Hair and Meat
- Per acre goats produce $350 opposed to cows at $125
- Pheasants readily adapt to life in confinement.
- Pheasants have a fast paced breeding rate.
- "Goats are the most underrated farm resource around today. If properly managed I would say this animal is about twice as profitable as cattle. Goats thrive in pasture land too poor to support cattle". Shelly Andrew, extension agent in Jackson County, NC.
- "Goats could be the savior of the US livestock industry". Wess Hallman, President Alabama Dairy Goat Association.
- "As ethnic diversity continues to grow in the US, demand for goat meat will increase as well". Ray McKinne, Animal Science Specialist at North Carolina A&T University.
- "I am very optimistic about the future of production of meat goats as the demand for goat meat, both domestic and foreign, is certain to increase". Clair E. Terrill, PH.D USDA

ARS

- "There are indications that the consumption of goats has moved up substantial since the mid-1980's at more or less stable prices". Frank Pinkerton, Extension Goat Specialist, Tuskegge University
- "Demand for cashmere, the fine underdown from cashmere goats that has long been preferred by royalty, has always exceeded supply". Goat Handbook, USDA
- "The demand for goat meat has continued to increase dramatically over the last decade and a half". The goat slaughter numbers pale in comparison to the slaughter numbers of the other red meat species, cattle and sheep. However, of the 3 only goat numbers have significantly increased over the last decade and a half. The other 2 have decreased or remained steady". Terry A. Gipson, PH.D Virginia State University.
- "It maybe awhile before goat meat is available in every market in America, but it is on its way". Merle Ellis, "The Nation's Best Known Butcher".
- "Over the past 15 years the game bird industry has grown tremendously. There seems to be an ever increasing demand for both game birds for restock and game birds for food." Bill McFarlane, McFarlane Pheasant Farm

2.4 OVERALL OBJECTIVE

2.4.1 Objectives

To become a viable and profitable small scale ranch. WA Ranch's objective is to start, maintain and increase the proposed ranching enterprise, thus increasing profit and providing for growth of the ranch over a 5 year period. After the 5 year period to remain a viable and profitable livestock producer.

Obtaining objectives include:

- Selling at the highest price for the year.
- Selling above the average price for the year.
- To make a profit.
- Selling at a price to meet cash flow needs.
- Concentrating on minimizing risk.

2.4.2 Strategy (long range) Goals.

- To become a viable small scale livestock producer.
- To establish and maintain a profitable sales margin.
- To develop new local marketing in roads for products.

3.0 MARKET ANALYSIS

3.1 Market Research

Specific target market:

Livestock Auctions

Retail Outlets - Upscale Stores, Healthfood Stores,

Restaurants, Caterers
Cooperatives
Hunting Preserves
Raise and Release Programs
Individuals - Breeding Stock, Personal Consumption

Overall Target Market
Ethnic Persons
Health Conscious Consumers
Upscale "Yuppie" Consumers & Establishments

3.2 Results of market research:

The target market for goat meat is divided into three parts.
Ethnic Groups - In some parts of the world goat is the preferred meat. Among ethnic groups from Africa, Caribbean, West Indies, Central America and the Middle East goat meat is a staple. It is also consumed by the Greek, Italian, Chinese and Korean. Goat (Cabrito) is a major part of the Hispanic diet. Hispanics now numbering more than 19 million and it is projected they will outnumber blacks as the nations largest minority group by 2015. Goat consumption is related to the Muslim religion. The Muslim population is around 14,000,000 and grow rapidly. Peoples from the Caribbean number slightly less than 2 million. The French consider goat (Chevon) a delicacy. Overall the increase in demand for goat meat appears to be good due to immigration which is averaging approximately 61,150 per month and will likely continue at that pace. Also, the economic status of ethnic groups preferring goat continues to improve.
Ethnic Restaurants - are the fastest growing segment of the food industry. In a recent study over a 4 year period restaurant patronage only increased 10% but rates at Mexican Restaurants increased by 43% and Asian restraints say a 54% increase. Chinese is the most popular of the restaurants followed by Mexican and Italian. All three incorporate goat meat into their diets.
Goat Meat is becoming popular with the health food sector and the "yuppie" sector and goat meat is now being consumed as gourmet item creating a niche market. Goat meat is higher in protein and iron and lower in fat and cholesterol than most meats. In 1991 Dr. Robert Degner, Professor and Director of the Florida Agricultural Market Research Center at the University of Florida in Gainsville surveyed 600 consumers to rate goat meat and beef and to rate a number of product characteristics. 25% had eaten goat meat before. For "overall appeal" the goat meat ratings showed 42% preferred the goat sample, 38% preferred beef and 20% could not distinguish between the two. The conclusions of the test where that properly prepared goat meat compares favorably with beef.

Goat meat demand has increased dramatically over the last decade. The USDA's figures reflect a 900% increase in a 16 year period.

Raising Goats

Goats are foragers and often grazed on land not suitable for other livestock. They love weeds and will eradicate Leafy Spurge, Tumble Weeds, Poison Ivy and many others. Only additional feedings are required in harsh winters or during pregnancy. Placing goats on a feedlot has proven to be unsuccessful as they obtain all their nutritional needs from the forage alone.
Goats have a high reproductive rate of a 120/175% kid crop.
6 - 12 goats per acre can successfully be placed as opposed to one cow.
Female goats can way from 100-115#. Male goats can weight as much as 250#
Cashmere goats have a reputation for being hardy and independent. Goats are as sensitive to all the diseases and parasites as other ruminants. However, grazing habits and inborn resistance have given goats a better advantage.

Cashmere Hair

The commercial market for cashmere has been around for centuries. Local Cashmere buyers and Cashmere Co-ops buy hair continuously and fair prices. Hair can also be sold to hand spinners.
The fiber was used to line and curtain the Arc of the Covenant of the old testament. Demand for this luxurious hair has always exceeded supply. Cashmere is largely imported, coming from China, Turkey, Afghanistan, Iraq, Iran, Kashmere, Australia and New Zealand. The U.S. market took a foot hold recently because of the political unrest in these countries. It causes the inventory of Cashmere to fluctuate wildly at times. Manufactures started looking to the U.S. to stabilize the inventory in the market and thus prices. Cashmere goats have a reputation for being hardy. Cashmere goats are a type not a breed with Cashmere is the secondary hair or undercoat. The hair begins to shed in late winter and early spring when it can be combed or sheered. The long fine down is used in knitted garments and the shorter down is used in woven fabrics. The separated guard hairs go into rugs and is used in tailored garments.

Pheasants

There are typically 4 major markets for pheasants. Gourmet food markets (upscale grocery stores, catering establishments and restaurants), private individuals that buy birds either for breeding or custom slaughter, individuals and groups that release the birds into the wild and hunting preserves.

Pheasants bred very well in captivity. They are easy to produce with the average hen producing 15 eggs. Incubation of the eggs takes about 23-25 days. Eggs are normally hatched in May. Pheasants are marketed between 22-24 weeks of age and are usually sold in September or October. Costs to begin a large scale pheasant are fairly low. The chicks costing anywhere from $1 to $1.50. Due to their large egg crop and adaptability to captivity a small amount of starter chicks can eventually produce a large flock.

Pheasant meat is suited to the healthy eating habits of the American population due to its national value. It is similar to turkey in fat and protein content and is moister than turkey and other wild meat.

There is a perception that pheasant is similar to lobster as a gourmet food item. And the price for pheasant often reflects that. Upscale restaurants and grocery stores often offer pheasant as a specialty food. More people are also becoming interested in purchasing dressed birds to cook in their own homes.

Many hunters belong to licensed hunting preserves. The popularity of these hunting preserves is on the rise due to the fact they tend to be located closer to the population and hunters don't have to travel as far for the sport. As the wild game bird population has dwindled more and more of these preserves have been founded and the need to purchase these birds has also increased.

3.4 COMPETITIVE FACTORS

3.4.1 Description and assessment of competitors

Judith Smith- Ft. Collins, CO

Judith raises a Boer Cashmere Goat. She successfully raises 150 goats on 10 acres. She shears and ships the cashmere hair to Cashmere America who in turn grade the hair and send her a check. She also, in conjunction with her goat producing neighbors, hires a livestock trucking company to haul the goats to Matins Stockyard in California. Her goats consistently bring $100 a head.

4.0 MARKETING

4.1 Marketing Orientation

4.1.1 Marketing Philosophy

WA Ranch will position itself as a local producer of specialty meats and game through extensive contact and marketing to it's specific target market. See section 3.1

5.0 SITUATION ANALYSIS

5.1 Resources

5.1.1 People

Owner - 15 Years of Experience in Administration, Marketing and

Management.

5.2 Opportunities

5.2.1 Industry Changes

- Growing market due to expanding ethnic immigration.
- More people eating at upscale establishments or buying from upscale "yuppie" grocers.
- More acceptance of goat meat.
- More people "health conscious" of fat intake.
- More organization of groups within the livestock industry. i.e. Cooperatives and Associations.

5.2.2 Problem Situations

- Not as accepted as red meat or poultry.
- No set marketing channels and not as organized getting livestock to market as cattle or sheep production.
- Livestock more susceptible to predator losses than larger livestock.

5.2.3 Company Weaknesses

- Management does not have lengthy experience in ranching.

5.4 MARKETING STRATGEY

Several local goat producers have expressed tremendous interest in developing a co-op to get the goats to market. WA Ranch will take the lead in this effort to organize these producers to come up with ways to maximize every producers profits by sharing expenses for goat transportation to market and to share "innovative" ideas regarding production and sales. Texas and California livestock auctions pay substantially more for goats than Colorado due to the larger ethnic population in each state and better coordination of the processing plants.

Plans to review weekly livestock pricing nationwide through Internet access to the Livestock Auction Reports and to stay up to date with the Goat and Pheasant Industry through Internet access to various groups and publications in the livestock arena.

Obtain names of hunting ranges and release programs from game commissioners.

Direct marketing literature and follow up calls to develop a relationship with local target market.

Self promotion and industry promotion through media relations.

6.0 FINANCIAL DATA

6.1 Current Financial Position

6.1.1 Startup costs

See section 1.5 for the startup costs and the projected monthly expenditures.

6.1.2 Profit and Loss Statement

Not applicable at this time

6.1.3 Cash Flow analysis

Not applicable at this time

6.2 PAYABLES/RECIVEBLES

6.2.1 Debts

Current mortgage on ranch of $66,500

6.2.1 RECEIVABLES

None at this time

6.3 FINANCIAL PROJECTIONS

See section 2.5.1 for general projections. Specific financial projects are not applicable at this time.

7.0 ORGANIZATION AND MANAGMENT

7.1 Key Personnel

Owner/Manager

Jackie Spilker, Matheson, CO 80303

Responsibilities

Owner - Marketing, Administration, Operations Management

8.0 OWNERSHIP

8.1 Structure of Business

8.1.1 Legal form of business:

WA Ranch is a sole proprietorship

9.0 CRITICAL RISKS

9.1 Description Of Risks

Year	1	2	3	4	5
Revenue Best	24,521	36,221	53,753	71,285	97,817
Revenue Most Likely	17,280	25,481	37,742	49,643	67,294
Revenue Worst	9,630	15,460	21,730	28,000	36,770

9.1.1 Sales Projection Not Attained

If sales projections are not attained, there are two possible outcomes: break even point is met or break even point is not met. If the break even is not met, then the company is not viable as it currently exists and will be modified. If the break even point is met, the projections and target market will be reevaluated.

9.1.2 Unforeseen Industry Trends

This is unknown at this time.

9.1.3 Unforeseen Economic Political and Social Developments

This is unknown at this time.

9.1.4 Capital Shortages

This is unknown at this time.

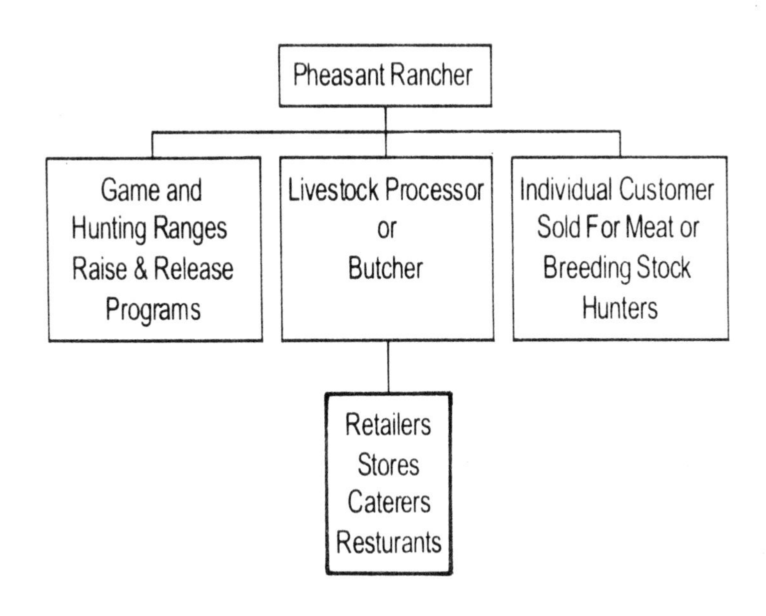

Pheasant Rancher
Game and Hunting Ranges Raise & Release Programs
Livestock Processor or Butcher
Individual Customer Sold For Meat or Breeding Stock Hunters
Retailers Stores Caterers Resturants

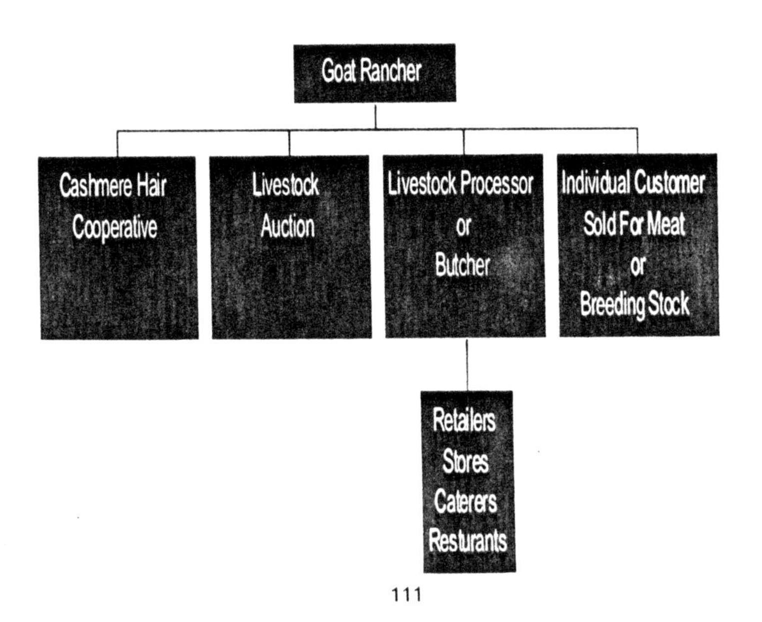

Goat Rancher
Cashmere Hair Cooperative
Livestock Auction
Livestock Processor or Butcher
Individual Customer Sold For Meat or Breeding Stock
Retailers Stores Caterers Resturants

CHAPTER 8 - FEDERAL HELP FOR FARMERS, RANCHERS AND THE AGRIUCLTURAL COMMUNITY

"I urge young farmers just starting out to look into these programs. It may be just the thing many of them need to start their own operations on the right track." Dan Glickman, USDA

Farm programs are created and administered by the United States Department Of Agriculture, or USDA for short, a branch of the Federal Government. Why farm programs? According to the USDA "Since the late 1920's, American farm policy has tried to encourage the production of adequate supplies of food and fiber and to maintain reasonable prices for consumers while, at the same time, assuring farmers a fair return on investment".

If you are a new farmer or rancher or are contemplating a farm enterprise you should familiarize yourself with the following programs:

Beginning Farmer/Rancher Program: Assists qualified beginning farmers and ranchers with operating and farm/ranch ownership loans. Administered by the Farm Service Agency.

Farm/Ranch Operating Loans: Available to individuals, partnerships, joint operations, cooperatives and corporations that are directly involved with farming and ranching on a family-sized farm. Administered by the Farm Service Agency.

Farm/Ranch Ownership Loans: Loans can be used to purchase or enlarge and existing ranch or farm. Also, to construct and improve buildings and to improve the environmental soundness of the farm/ranch. Administered by the Farm Service Agency.

Farm Ownership Downpayment Loans: Direct loan for up to 30% of the purchase price of a family sized farm. Administered by the Farm Service Agency.

Rural Youth Loans: Loans made to individuals, up to $5,000, who are sponsored by a project manager such as 4-H, FFA or local instructor. Must be 10-20 years of age. Administered by the Farm Service Agency.

Limited Resource and Socially Disadvantaged Farm Loans: Farm/ranch loans allocated for operations and ownership. Limited resource and socially disadvantaged includes: beginning farmers, low income farmers/ranchers, minorities and women. Administered by the Farm Service Agency.

Soil,Water and Wildlife Loans and Direct Payments: Loans to farmers and land owners for the purpose of developing, conserving and making proper use of land and water resources and wildlife habitats. If you own land and do not intend to use it for farming the government will pay you to keep "environmentally sensitive" land out of farming and ranching. Administered by the Natural Resources Conservation Service.

Agricultural Grants: Can you build a better chicken? Find a new way to process food? Individuals and Businesses demonstrating expertise in agricultural areas

can apply for some of the millions of dollars set aside for Agricultural Research Grants: Small Business Innovation Research Grants; Grants For Agricultural Research, Special Research Grants; Grants For Agricultural Research, Competitive Research Grants; Sustainable Agriculture Research and Education; Alternative Agricultural Research and Commercialization Program.
Administered by: Cooperative State Research, Education and Extension Service, USDA, AG Box 2201, Washington, DC 20250-2201 (202)720-4423

Federal Farm and Agricultural Related programs:

FARM OPERATING LOANS

Agency: Farm Service Agency

Range & Average of Financial Assistance: Direct loans up to $200,000; guaranteed loans up to $400,000; Direct average loan size about $43,000 and guaranteed average loan size about $115,000 for fiscal year 1996.

Objective: To enable operators of not larger than family farms through the extension of credit and supervisory assistance, to make efficient use of their land, labor, and other resources, and to establish and maintain financially viable farming and ranching operations.

Type of Assistance: Direct Loans; Guaranteed/Insured Loans.

Summary:Applicants/borrowers are the direct beneficiaries and must meet the applicant eligibility requirements. Families, individuals, and entities who are farmers, ranchers or aquaculture operators are the beneficiaries. Loan funds may be used to: (1) Purchase livestock, poultry, fur bearing and other farm animals, fish, and bees; (2) purchase farm, forestry, equipment; (3) provide operating expenses for farm, forestry, enterprise; (4) meet family subsistence needs and purchase essential home equipment; (5) make minor real estate improvements; (6) refinance secured and unsecured debts subject to certain restrictions; (7) pay property taxes; (8) pay insurance premiums on real estate and personal property; (9) finance youth projects; (10) plant softwood timber on marginal land; (11) subordination's to borrowers to enable them to obtain annual operating credit from another lending source; and (12) support other miscellaneous purposes. Use restrictions are shown under Applicant Eligibility.

Application Process: Applicants file Form FmHA 410-1, Application for FSA Services, with supporting information, at the local county office of the Farm Service Agency for direct loans or Form FmHA 1980-25 with the prospective lender for loan guarantees.

FARM OWNERSHIP LOANS

Agency: Farm Service Agency

Range & Average of Financial Assistance: Maximum direct $200,000, maximum guaranteed $300,000. Average direct $80,000, guaranteed $171,000.

Objective: To assist eligible farmers, ranchers, and aquaculture operators, including farming cooperatives, corporations, partnerships, and joint operations, through the extension of credit and supervisory assistance to: Become owner-operators of not larger than family farms; make efficient use of the land, labor, and other resources; carry on sound and successful farming operations; and enable farm families to have a reasonable standard of living.

Type of Assistance: Direct Loans; Guaranteed/Insured Loans.

Summary:Applicants/borrowers are the direct beneficiaries and must meet the applicant eligibility requirements. Families, individuals, and entities who are farmers, ranchers or aquaculture operators are the beneficiaries. Loan funds may be used to: (1) Enlarge, improve, and buy family farms; (2) provide necessary water and water facilities; (3) provide basic soil treatment and land conservation measures; (4) construct, repair, and improve essential buildings needed in the operation of a family farm; (5) construct or repair farm dwellings; (6) improve, establish, or buy a farm-forest enterprise; (7) provide facilities to produce fish under controlled conditions; (8) acquire farmland by socially disadvantaged individuals who will be provided the technical assistance necessary in applying for direct farm ownership (FO) loan.

Application Process: Applicants file Form FmHA 410-1, Application for FSA Services, with supporting information, at the local county office of the Farm Service Agency for direct loans or Form FmHA 1980-25 with the prospective lender for loan guarantees.

INTEREST ASSISTANCE PROGRAM

Agency: Farm Service Agency, USDA

Range & Average of Financial Assistance: $1 to $400,000; $121,000

Objectives: To aid not larger than family sized farms in obtaining credit when they are temporarily unable to project a positive cash flow without a reduction in the interest rate.

Type of Assistance: Guaranteed/Insured Loans

Summary:Interest Assistance Program can be used on any of the three types of guaranteed loans. The three types and loan purposes are as follows: (1) Farm Ownership (FO) Loans - to buy, improve, or enlarge farms. Uses may include construction, improvement, or repair of farm homes and service buildings; improvement of on-farm water supplies; to help farmers supplement their farm income; (2) Operating Loans (OL) - to pay for items needed for farm operations, including livestock, farm and home equipment, feed, seed, fertilizer, fuel, chemicals, hail and other crop insurance, family living expenses, minor building improvements, water system development, hired labor, and methods of operation to comply with the Occupational Safety and Health Act. The loan limit is $300,000 for FO and $400,000 for OL.

Application Process: Application for a guaranteed loan with interest assistance can be made by contacting a lender for or further information on procedures, forms and requirements for making an application can be obtained from the FSA county office in the county where the proposed farming operation or headquarters will be located. FSA has county offices serving every rural county in the United States. Farm Service Agency offices are listed in the phone directory under U.S. Government, Department of Agriculture. Location of an office may be obtained by writing to United States Department of Agriculture, Farm Service Agency, Washington, DC 20250.

EMERGENCY LOANS

Agency: Farm Service Agency; USDA

Range & Average of Financial Assistance: $500 to $500,000; $58,000

Objective: To assist established (owner or tenant) family farmers, ranchers and aquaculture operators with loans to cover losses resulting from major and/or natural disasters, which can be used for annual farm operating expenses, and for other essential needs necessary to return disaster victims' farming operations to a financially sound basis in order that they will be able to return to private sources of credit as soon as possible.

Type of Assistance: Direct Loans.

Summary:Applicants/borrowers are the direct beneficiaries when they meet all eligibility criteria. Families, individuals and entities who are farmers, ranchers or aquaculture operators are the beneficiaries. Loan funds may be used to repair, restore, or replace damaged or destroyed farm property (real and chattel) and supplies which were lost or damaged as a direct result of a natural disaster; under certain conditions, refinance secured and unsecured debts made necessary by the disasters; finance adjustments in the farming, ranching or aquaculture operation(s) determined necessary to restore or maintain applicants' operations on a sound financial basis equivalent to their predisaster potential. The total of all actual loss loans is based on actual dollar value of production and physical losses. Loans are made at 3.75% interest, with a maximum limit of 80% of the actual production loss and 100% of the actual physical loss, or $500,000 total indebtedness. Loans are made in counties: (1) Named by the Federal Emergency Management Agency as being eligible for Federal assistance under a major disaster or emergency declaration by the President; (2) designated as natural disaster areas by the Secretary of Agriculture; and (3) designated by the FSA Administrator for severe physical losses, only, as a result of a natural disaster.

Application Process: Application Form FmHA 410-1 provided by the Farm Service Agency must be presented, with supporting information, to the FSA county office serving the applicant's county. FSA personnel assist applicants in completing their application forms.

FARM LABOR HOUSING LOANS AND GRANTS

Agency: Rural Housing Service, USDA

Range & Average of Financial Assistance: Initial grants $130,000 to $2,300,000; $1,104,120. Initial loans to individuals $20,000 to $200,000; $34,500. Initial loans to organizations $165,000 to $670,000; $292,753.

Objective: To provide decent, safe, and sanitary low-rent housing and related facilities for domestic farm laborers.

Type of Assistance: Project Grants; Guaranteed/Insured Loans.

Summary:Loans are available to farmers, family farm partnership, family farm corporation, or an association of farmers. Loans and grants are available to States, Puerto Rico, the U.S. Virgin Islands, political subdivisions of States, broad-based public or private nonprofit organizations, federally recognized Indian Tribes and non- profit corporations of farm workers. Grants are available to eligible applicants only when there is a pressing need and when it is doubtful that such facilities could be provided unless grant assistance is available. The loans and grants may be used for construction, repair, or purchase of year-round or seasonal housing; acquiring the necessary land and making improvements on land for housing; and developing related support facilities including central cooking and dining facilities, small infirmaries, laundry facilities, day care centers, other essential equipment and facilities or recreation areas. Funds may also be used to pay certain fees and interest inci- dental to the project. Restrictions on the use of funds are: The housing must be of a practical type and must be constructed in an economical manner and not of elaborate material or extravagant design. Housing financed with labor housing loan or grant funds must be occupied by domestic farm laborers and their families.

Application Process: Preapplication will be made on Form SF 424.2 "Application for Federal Assistance," and submitted to the District/State offices of the Rural Development Services (formerly Farmers Home Administration) fully documenting the need for the loan or grant and the proposed amount needed. Attachments relating to the size of the proposed project, estimated cost, budget and need are also required. This program is subject to the provisions of 7 CFR 3015 or 3016.

EMERGENCY CONSERVATION PROGRAMS (ECP)

Agency: Farm Service Agency

Range & Average of Financial Assistance: $50 to $64,000; $2,681

Objectives: To enable farmers to perform emergency conservation measures to control wind erosion on farmlands, or to rehabilitate farmlands damaged by wind erosion, floods, hurricanes or other natural disasters and to carry out emergency water conservation or water enhancing measures during periods of severe drought.

Type of Assistance: Direct payments for specified use.

Summary:Any person who as owner, landlord, tenant or sharecropper on a farm or ranch, including associated groups, and bears a part of the cost of an approved conservation practice in a disaster area, is eligible to apply for cost share conservation assistance.

Application Process: Eligible persons make application on Form ACP-245, for cost sharing, at the country FSA office for the county in which the land is located.

WATER BANK PROGRAM

Agency: Natural Resources Conservation Service, USDA

Range & Average of Financial Assistance: $7 to $75 per acre; $13.00

Objectives: To conserve surface waters, preserve and improve the Nation's Wetlands, increase migratory waterfowl habitat in nesting, breeding and feeding in the U.S. and secure environmental benefits for the Nation.

Type of Assistance: Direct payments for specified use.

Summary:Landowners and operators of specified types of wetlands in designated

important migratory waterfowl nesting, breeding and feeding areas.
Agreements are for 10 years with eligible landowners to preserve important nesting, breeding and feeding areas of migratory waterfowl. During the agreement the participants agree in return for annual payments not to drain,
burn, fill or otherwise destroy the wetland character of such areas and not to use areas for agricultural purposes.

Application Process: Make application at the NRCS office for the county in which the land is located. Applicants must develop an approved conservation plan with local soil and water conservation district representatives.

FORESTRY INCENTIVES PROGRAM (FIP)

Agency: Natural Resource Conservation Service, USDA

Range & Average of Financial Assistance: $50 to $10,000 per year, $1,600

Objective: To bring private non-industrial forest land under intensified management, to increase timber production, to assure adequate supplies of timer, and to enhance other forest resources thorough a combination of public and private investments on the most productive sites on eligible individual or consolidated ownership of efficient size and operation.

Type of Assistance: Direct payment for specified use.

Summary:A private individual, group, association, Indian Tribe or other native group, corporation (except corporations whose stocks are publicly traded) or other legal entity which owns "non-industrial" private forest lands capable of producing industrial wood crops is eligible to apply for cost-sharing assistance. Cost-share agreement are limited to eligible ownership of land of not more than 1,000 acres of non-industrial private forest land, capable of producing at least 50 cubic feet of wood per acre per year, except by special approval. Cost-sharing of up to 65% of the total cost is available under the Forestry Incentives Program for tree planting, timber stand improvement, and site preparation for
natural regeneration.

Application Process: Eligible persons will make application for annual cost-sharing agreements. Applications may be filed at any time during the year at the NRCS office for the designated county in which the land is located.

CONSERVATION RESERVE PROGRAM (CRP)

Agency: Farm Service Agency, USDA

Range & Average of Financial Assistance: $50 to $50,000; $5,324

Objective: To protect the Nation's long-term capability to produce food and fiber; to reduce soil erosion; to reduce sedimentation; to improve water quality; to
create a better habitat for wildlife; to provide some needed income support for farmers.

Type of Assistance: Direct Payments for Specified Use.

Summary:An individual, partnership, association, Indian Tribal ventures, corporation, estate, trust, other business enterprises or other legal entities and, whenever applicable, a State, political subdivision of State, or any agency thereof owning or operating private croplands, and State or local government croplands will benefit. Eligible owners or operators may place highly erodible or other environmentally sensitive land into a 10-15 year contract. The participant, in return for annual payments, agrees to implement a conservation plan approved by the local conservation district for converting highly erodible cropland or other environmentally sensitive land to a long-term resource conserving cover i.e., eligible land must be planted with a vegetative cover, such as, perennial grasses, legumes, forbs, shrubs, or trees. The participant agrees to reduce the aggregate total of allotments and quotas for the contract period for each farm which contains land that is subject to a Conservation Reserve Program contract by an amount based on the ratio of the total cropland acreage on each farm, to the total acreage on each farm subject to the CRP contract. Financial and technical assistance are available to participants to assist in the establishment of a long-term resource conserving cover.

Application Process: FSA has two methods for enrolling acreage in the CRP. One method is a continuous sign-up process where acreage suitable for certain environmental priority practices, including but not limited to grass waterways, riparian buffers or filterstrips, and acreage within wellhead protection areas may be offered and accepted without going through a competitive offer process. The second method is to offer acreage during a general sign-up period where offers to enroll highly erodible and other environmentally sensitive acreage are accepted and ranked competitively based on an environmental benefits index. Regardless of the method of enrollment, the local FSA office, that serves the area in which the farm or ranch is located, will provide applicants the maximum payment rate CCC will accept to enroll certain acreage in the program. Applicants will submit a rental rate per acre offered that may be equal to or less than CCC's maximum payment rate.

GREAT PLAINS CONSERVATION

Agency: Natural Resources Conservation Service, USDA

Range & Average of Financial Assistance: Up to $35,000 per farm operating unit over a contract period running from 3 to 10 years.

Objective: To conserve and develop the Great Plains soil and water resources by providing technical and financial assistance to farmers, ranchers, and others in planning and implementing conservation practices.

Type of Assistance: Direct Payments for Specified Use; Advisory Services and Counseling

Summary: Applicants must have control of the land for the period of the contract running from a minimum of 3 years to a maximum of 10 years. Applicants must be in compliance with Title XIV, Subtitles A and B of Public Law 101-624. Land must be located in one of 556 designated counties within the States of Colorado, Kansas, Montana, Nebraska, New Mexico, North Dakota, Oklahoma, South Dakota, Texas and Wyoming. Cost-share funds are available for many of the soil and water conservation measures determined to be needed to protect and stabilize a farm or ranch unit against climatic and erosion hazards of the Great Plains area, and applied in accordance with a conservation plan for the entire operating unit. Additional practices may be included in the plan for agriculture-related pollution abatement, enhancement of fish, wildlife, recreational resources, and promotion of economic use of land and may or may not be cost-shared.

Application Process: Application is made to the Natural Resources Conservation Service district conservationist serving the county in which the operating unit is located. The application is submitted on Form SCS-LTP-001, "Application for long-term contracted assistance through the Great Plains Conservation Program."

WILDLIFE HABITAT INCENTIVE PROGRAM (WHIP)

Agency: Natural Resources Conservation Service, USDA

Range & Average of Financial Assistance: Cost-share payments are generally limited to $10,000 per contract. Average contract payments are estimated to be $2,000.

Objective: This program was created to develop upland wildlife habitat, wetland wildlife habitat, threatened and endangered species habitat, fish habitat and other types of wildlife habitat.

Type of Assistance: Direct Payments for Specified Use.

Summary: A participant may be an owner, landlord, operator, or tenant of eligible lands. Limited resource producers, small scale producers, producers of minority groups, Federally Recognized Indian Tribal Governments, Alaska natives, and Pacific Islanders are encouraged to apply. Technical assistance is provided to develop a Wildlife Habitat Development Plan for eligible participants.
Cost-share payments may be made to implement wildlife habitat practices. 75% cost-share is available to reimburse participants for installing practices beneficial to wildlife.

Application Process: Program participation is voluntary. The applicant applies at the local USDA service center on Form CCC-1200. Applications may be filed at any time during the year. The participant develops and submits a Wildlife Habitat Development Plan for the land unit of concern. The participant's plan serves as the basis for the WHIP contract. Technical assistance, and cost-share payments may be provided to apply the needed wildlife habitat practices within a time schedule specified by the plan. A contract with a participant may apply more than one wildlife habitat practice.

ENVIRONMENTAL QUALITY INCENTIVES PROGRAM (EQIP)

Agency: Natural Resources Conservation Service, USDA

Range & Average of Financial Assistance: Cost-share and incentive payments are limited to $10,000 per person per year and to $50,000 over the length of the contract. Average contract payments are estimated to be $30,000.

Objectives: Individual/family farmers and ranchers who face serious threats to soil, water, and related natural resources, or who need assistance with complying with Federal and State environment laws. A participant may be an owner, landlord, operator, or tenant of eligible agricultural lands. Limited resource producers, small-scale producers, producers of minority groups, Federally recognized Indian tribal governments, Alaska natives, and Pacific Islanders are encouraged to apply.

Type of Assistance: Direct Payments for Specified Use.

Summary: Individual/family farmers and ranchers who face serious threats to soil, water, and related natural resources, or who need assistance with complying with Federal and State environmental laws. A participant may be an owner, landlord, operator, or tenant of eligible agricultural lands. Limited resource producers, small-scale producers, producers of minority groups, Federally Recognized Indian Tribal Governments, Alaskanatives, and Pacific Islanders are encouraged to apply. Technical assistance is provided in conservation planning for eligible participants. Education and financial assistance is provided for implementation of structural, vegetative, and land management practices. Cost-share payments may be made to implement one or more eligible structural or vegetative practices. Incentive payments can be made to implement one or more land management practices. Fifty percent of the funding available for technical, cost-share payments, incentive payments, and education shall be targeted at practices relating to livestock production.

Application Process: Program participation is voluntary. The applicant applies at the local USDA Service Center on Form CCC-1200. Applications may be filed at any time during the year. The participant develops a conservation plan for the farm or ranching unit of concern. The participant's conservation plan serves as the basis for the EQIP contract. Technical assistance, educational assistance and cost-share or incentive payments may be provided to apply needed conservation practices and land use adjustments within a time schedule specified by the conservation plan. A contract with a participant may apply one or more land management practices or one or more structural or vegetative practices or both.

WETLANDS RESERVE PROGRAM (WRP)

Agency: Natural Resources Conservation Service

Range & Average of Financial Assistance: NA

Objective: To restore and protect farmed wetlands, prior converted wetlands, wetlands farmed under natural condition, riparian areas, and eligible buffer areas for landowners who have eligible land on which they agree to enter into a permanent or long-term easement or restoration agreement contract with the Secretary. The goal of WRP is to have 975,000 acres enrolled by the year 2000
with one-third as permanent easements, one-third as 30-year easements and one-third as restoration agreement acres.

Type of Assistance: Direct Payments for Specified Use.

Summary: An individual landowner, partnership, association, corporation, estate, trust, other business enterprises or other legal entities and, whenever applicable, a State, a political subdivision of a State; or any agency thereof owning private croplands will benefit. Eligible landowners may offer farmed wetlands, prior converted wetlands, wetlands farmed under natural condition, intensively managed pasture and hayland riparian areas, along with eligible buffer areas to be placed under a permanent or 30-year easement or
restoration agreement. A deed restriction covering the land approved under easement must be recorded in the local land deeds office. The landowner will receive financial and technical assistance to install necessary restoration practices on the land under easement or the practice will be installed
by the Secretary. Subject to the acceptance of an offer by the Federal Government, the landowner will receive in cash an amount specified in the WRP contract but not to exceed the fair agricultural market value of the land "as is" condition less the fair market value of such land encumbered by the permanent easement or 75% for a 30-year easement. The landowner shall ensure that the easement granted to Natural Resources Conservation Service (NRCS) is superior to the rights of all others and shall agree to

implement a wetland restoration plan designed to restore and maintain the easement area. The plan
will include a designated access route to be used as necessary for easement management and monitoring. The landowner shall agree to a permanent retirement of crop acreage bases, allotments, and quotas to the extent that the sum of the crop acreage bases and allotments will not exceed the remaining cropland of the present farm or subsequently reconstituted farm. In cases
involving restoration agreements there is only on restoration cost-share and no land payment.

Application Process: Submit an intention to enroll to the local NRCS office that serves the area in which the farm or ranch is located during the designated sign-up period.

CROP INSURANCE

Agency: Risk Management Agency, USDA

Range & Average of Financial Assistance: Level of assistance varies according to policy, crop and indemnities paid.

Objective: To promote the national welfare by improving the economic stability of agriculture through a sound system of crop insurance and providing the means for the research and experience helpful in devising and establishing such insurance.

Type of Assistance: Insurance.

Summary:Any owner or operator of farmland, who has an insurable interest in a crop in a county where insurance is offered on that crop is eligible for insurance. Producers will be covered under the Noninsured Assistance Program (NAP) which is available to provide coverage equivalent to the catastrophic risk protection in areas where catastrophic risk protection is not available, if such crop is produced for food or fiber and the area is authorized. The Federal Crop Insurance Corporation (FCIC) is a wholly owned government Corporation that provides crop insurance to crop producers against losses because of unavoidable causes and/or uncontrollable events. The Federal Crop Insurance Reform Act of 1994 made significant changes in the program to provide more of the nation's producers with an ongoing source of risk protection to reduce the need for ad hoc disaster payment assistance. The Federal Agriculture Improvement and Reform Act of 1996 provided for more changes, including the provision for a more extensive risk management education program to assist and train producers on risk management strategies, including futures and options trading and insurance protection programs. Catastrophic crop insurance protection is fully subsidized except for a minor processing fee to be
paid by the farmer. This coverage compensates the producer for yield losses exceeding 50% of yield and at a price equal to 60% of maximum price. Additional protection at higher level of coverage will continue to be offered. Coverage levels will be subsidized to the extent of the premium on at least the catastrophic level of coverage and the delivery costs. The Noninsured Crop Disaster Assistance Program, which the Farm Service Agency now administers, is available to provide coverage equivalent to the catastrophic crop insurance protection in areas where catastrophic crop insurance protection is not available and if such crop is produced for food or fiber. There are several pilot insurance programs in limited areas for 1997. Included in these pilot programs are the Income Protection (IP) program, the Revenue Assurance (RA) program, Group Risk Plan (GRP) of Insurance, and Crop Revenue Coverage (CRC). Insurance is offered on the following crops and/or commodities, with specific crops being insured under the new

pilot programs: Almonds, apples, fresh apricots, processing apricots, avocados, avocado trees, barley (feed, malting), dry beans, beans canning/processing), blueberries, canola, carambola trees, citrus, citrus trees, corn for silage/production, hybrid cord see, cotton, EIS cotton, cranberries, figs, flax, forage production, forage seeding, grain sorghum, grapes, table grapes, macadamia nuts, macadamia trees, mango trees, millet, nectarines (fresh market), nursery stock, oats, onions, peaches (fresh market/processing), peanuts, pears, dry peas, green peas, peppers, plum (fresh market), popcorn, potatoes, prunes, raisins, rice, rye, safflower, hybrid sorghum seed, soybeans, sugar beets, sugarcane, sunflowers, sweet corn (canning/freezing), sweet corn (fresh market), tangelos, tobacco (guaranteed/quota), tomatoes (canning/processing), tomatoes (fresh market), walnuts, and wheat. Specific crops offered under the IP pilot program are avocados, corn, cotton, grain sorghum, soybeans, and wheat. Crops offered for the RA program are corn and soybeans. Barley, corn, cotton, forage, grain sorghum, peanuts, soybeans, and wheat are offered in the CRC program. Barley, corn, cotton, forage, grain sorghum, peanuts, soybeans, and wheat are offered under GRP.

Application Process: Form FCI-12, Application for Federal Crop Insurance for 1993 and succeeding crop years, or an application for multiple peril crop insurance offered by a company reinsured by FCIC, must be filed with the county FSA office or a crop insurance sales agent. Additional coverage is available only from private companies but the application is identical. Planted crops and acreage must be reported to establish eligibility for NAP.

FARM LAND PROTECTION PROGRAM

Agency: Natural Resources Conservation Service, USDA

Range & Average of Financial Assistance: FY 96 from $100,000 to $1,900,000 per cooperating entity; $387,162

Objective: To purchase conservation easements or other interests in lands to limit non-agricultural uses of farmland with prime, unique, or other productive soils.

Type of Assistance: Direct Payments for Specified Use.

Summary: Provides funds to help purchase development rights to keep productive farmland in agricultural uses. Working through existing programs, the USDA joins with state, tribal and local governments to acquire conservation easements or other interests from landowners. USDA provides up to 50% of the fair market easement value. To qualify farmland must be part of a pending offer from a State, tribe or local farmland protection program, be privately owned, have a conservation plan, be large enough to sustain agricultural production, be accessible to markets for what the land produces, have adequate infrastructure and agricultural support services, and have surrounding parcels of land that can support long term agricultural production.

Application Process: Application information is included in the Notice of Request for Proposals published in the Federal Register. It is available from the State Offices of the (NRCS), the Federal Register and USDA NRCS home page, and the Farm Land Information Center.

COMMODITY LOANS AND PURCHASES (PRICE SUPPORTS)

Agency: Farm Service Agency

Range & Average of Financial Assistance: Range and average not available. Loans $50 to $76,000,000; $24,288

Objectives: To improve and stabilize farm income, to assist in bringing about a better balance between supply and demand of the commodities, and to assist farms in the orderly marketing of their crops.

Type of Assistance: Direct payments with Unrestricted Use; Direct Loans

Summary: Owner, landlord, tenant or sharecropper on an eligible farm that has produced the eligible commodities, meets program requirements as announced by the Secretary and maintains beneficial interest in the commodity. Eligible commodities include feed grains, wheat, rice, peanuts, tobacco, dairy products (purchases only), upland cotton, extra long staple cotton, sugar, soybeans, canola, flaxseed, mustard seed, rapeseed, safflower, and sunflower seed. Instead of immediately selling the crop after harvest, a farmer who grows one or more of most field crops can store the produce and take out a non-recourse" loan for its value, pledging the crop itself as collateral. Nonrecourse means that the producer can discharge debts in full by forfeiting or delivering the commodity to the government. The loan allows farmers to pay their bills and other loan payments when they come due, without having to sell crops at a time of year when the prices tend to be at their lowest. Later, when the market conditions are more favorable, farmers can sell crops and repay the loan with the proceeds.

Application Process: In the case of warehouse-stored commodities, producer or cooperative marketing association presents warehouse receipts to the FSA county office. In the case of farm-stored commodities, producer/processor or cooperative or cooperative marketing association request a loan at the FSA county office.

PRODUCTION FLEXIBILITY PAYMENTS FOR CONTRACT COMMODITIES (CONTRACT COMMODITY DIRECT PAYMENTS)

Agency: Farm Service Agency, USDA

Range & Average of Financial Assistance: The production flexibility contract payments for the 1996 crop as of February 28, 1997 consisted of: $2,090,549,958 for feed grains; $1,939,501,369 for wheat; $698,450,136 for upland cotton, and $454,660,169 for rice. The production flexibility contract advance payments for the 1997 crop as of February 28, 1997 consisted of:
$2,180,208,476 for feed grains, $641,622,683 for wheat; $267,114,011 for upland cotton, and $215,535,851 for rice. Cotton, feed grain, wheat and rice production flexibility contract payments, in total, may not exceed $40,000 to any one person during any fiscal year. Annual crop year limits on the sum of marketing assistance gains and loan deficiency payments are continued at $75,000 per person for each of the 1996 through 2002 crops.

Objective: To support farming certainty and flexibility while ensuring continued compliance with farm conservation and wetland protection requirements.

Type of Assistance: Direct payments with unrestricted use.

Summary:Owner, landlord, tenant, or share cropper on a farm with contract acreage that meets program requirements as announced by the Secretary. Producers enrolled in the 7-year Production Flexibility Contracts during
the one-time sign-up held in 1996 are eligible to receive contract payments. All contracts, except those executed after the expiration of Conservation Reserve Program contracts (with an associated crop acreage base reduction), began with the 1996 crop and extend through the 2002 crop. A farm was eligible for enrollment if it had a wheat, corn, grain sorghum, barley, oats, upland cotton, or rice crop acreage base established for 1996. Once the farm is enrolled, the crop acreage base becomes contract acreage. Commodity-specific contract payment rates are determined annually based on the statutory spending levels and the amount of enrolled contract acreage. Farm level commodity payments are equal to the contract payment rate multiplied by 85% of the contract acreage multiplied
by the farm program payment yield. To be eligible for contract payments producers are required to: (a) comply with the conservation and wetland protection requirements on all of the producer's farms; (b) comply with planting
flexibility requirements; (c) use the contract acreage for an agricultural or related activity; and (d) obtain at least the catastrophic level of crop insurance for
each crop of economic significance or provide a written statement that waives any eligibility for emergency crop loss assistance; and (e) file annual acreage reports on any fruit or vegetable plantings on contract acreage. Annual payments will be made no later than September 30 of each of fiscal years 1997-2002. Producers may elect to receive 50% advance payments on December 15
or January 15 of the respective fiscal year.

Application Process: The farm operator visits the FSA office to sign Form-478, a Production Flexibility Contract and to report fruit and vegetable acreage planted for harvest on contract acreage on Form-578. This program is excluded from coverage under OMB Circular Nos. A-102 and A-110.

ALTERNATIVE AGRIUCULTURAL RESEARCH AND COMMERCIALIZATION PROGAM (AARC CENTER)

Agency: Alternative Agricultural Research and Commercialization (AARC) Center, USDA

Range & Average of Financial Assistance: $10,000 to $1,000,000;- $250,000

Objectives: To search for new non-food, non-feed products that may be produced from agricultural commodities and for processes to produce such products. To conduct product and co-product/process development and demonstration projects, as well as provide commercialization assistance for industrial products from agricultural and forestry materials and animal by products. To encourage cooperative development and marketing efforts among manufacturers, private and government laboratories, universities, and financiers to assist in bridging the gap between research results and marketable, competitive products and processes. To collect and disseminate information about commercialization projects that use agricultural or forestry materials and industrial products derived therefrom.

Type of Assistance: Cooperative Agreements.

Summary:Public and private educational and research institutions and organizations, Federal agencies, and individuals. Preference is given to private firms which will operate in or near rural areas or rural communities. Pre- proposals/proposals should focus

on products/processes from the following material categories: Starches/carbohydrates, fats and oils, fibers, forest materials, animal products, other plant materials uses as pharmaceutical, fine chemicals, encapsulation agents, rubber, etc. Primary interest is in providing assistance to technology development projects that will commercialize new industrial (non-food, non-feed) uses from new and existing agricultural and forestry materials. Special emphasis will be given to those pre-proposals/proposals whose products are closest to commercialization. Pre-proposals/proposals that request funds for research may be considered;
however, such requests must include an overall development plan that contains potential markets, development costs, and industry participation. No grant or cooperative agreement may be entered into for the acquisition or construction of a building or facility. Not more than 25% of the funds obligated each fiscal year shall be awarded only for projects concerning new products derived from animal sources.

Application Process: Solicitations are published in the Federal Register and in the Commerce Business Daily. Applications should be submitted to the Alternative Agricultural Research and Commercialization Center (AARC), USDA as outlined in the guidelines. Application procedures are provided by the AARC Center.

SMALL BUSINESS INNOVATION RESEARCH

Agency: Cooperative State Research, Education and Extension Service, USDA

Range & Average of Financial Assistance: $46,000 to $225,000; $94,990

Objective: To stimulate technological innovation in the private sector, strengthen the role of small businesses in meeting Federal research and development needs, increase private sector commercialization of innovations derived from USDA-supported research and development efforts, and foster and encourage participation, by women-owned and socially disadvantaged small business firms in technological innovation.

Type of Assistance: Project Grants

Summary: Small businesses which: (a) Are organized for profit, independently owned or operated, are not dominant in the proposed research field, have their principal places of business located in the United States, have a number of employees not exceeding 500 in all affiliated firms owned or controlled by a single parent concern, and meet the other regulatory requirements outlined in 13 CFR Part 121, as amended; (b) are at least 51% owned, or in the case of a publicly owned business, at least 51% of its voting stock is owned, by U.S. citizens or lawfully admitted permanent resident aliens; (c) are the primary source of employment for the principal investigator of the proposed effort at the time of award and during the actual conduct of proposed research; (d) are the primary performer of the proposed research effort. Because this program is intended to increase the use of small business firms in Federal research or research and development, the term "primary performer" means that a minimum of two-thirds of the research or analytical work, as determined by budget expenditures, must be performed by the proposing organization under phase I grants. For phase II awards, a minimum of one-half of the research or analytical effort must be conducted by the proposing firm. The selected areas for research are forests and related resources; plant production and protection; animal production and protection; air, water, and soils; food science and nutrition; rural and community development; aquaculture; industrial applications; and marketing and trade.

Application Process: Publication by the Small Business Administration listing all agencies participating in the program, their Small Business Innovation Research coordinators, proposed dates for their solicitations, and proposed topic areas. Formal proposal to SBIR Program, CSREES, USDA, as outlined in the SBIR program solicitation. Application procedures are contained in the SBIR program solicitation.

More Federal Farm Help

AgriAbility Program

Administered by the Cooperative Extension System. The program was created to help people with serious injury or permanent disabilities to farm or ranch. The program provides technical assistance and community awareness through case management.

Conservation Farm Option (CFO)

Pilot program through the Natural Resources Conservation Service for producers of wheat, feed grains, cotton and rice. Purposes of the program include: conservation of soil, water quality protection and improvement, wetland restoration, protection and creation, wildlife habitat development and protection and similar purposes. Eligibility is limited to owners and producers who have contract acreage enrolled in Production Flexibility Contracts. The CFO is a voluntary and participants are required to develop and implement a conservation farm plan. The CFO plan spans a 10 year period and is not restricted as to what measures may be included in the conservation plan as long as it provides environmental benefit. During the contract period the owner or producers receives annual payments for implementing the CFO contract and agrees to forgo payments under the CRP, Wetlands Reserve Program and the EQIP program in exchange for one consolidated program.

Conservation of Private Grazing Land Initiative (CPLG)

The CPLG will be administered by the Natural Resources Conservation Service and provide the technical, educational and related assistance to those who own private grazing lands. This is not a cost share program. The technical assistance will offer opportunities for better grazing land management, protect soil from erosive wind and water, use more energy efficient ways to produce food and fiber, conserve water, provide habitat for wildlife, sustain forage and grazing plants, use plants to sequester greenhouse gases, increase soil organic matter and use grazing lands as a source of biomass energy and raw materials for industrial products.

Flood Risk Reduction Program (FRR)

The FRR was established to allow farmers who voluntarily enter into contracts to receive payments on land with high flood potential. In return, participants agree to forego certain USDA program benefits. These contract payments provide incentives to move farming operation from frequently flooded land. Contact the Farm Service Agency.

Stewardship Incentives Program (SIP)

The SIP administered through the Forest Service provides technical and financial assistance to encourage non-industrial private forest landowners to keep their lands and natural resources productive and healthy., Qualifying land includes rural lands with exiting tree cover or land suitable for growing trees and which is owned by a private individual, group, association, corporation, tribe or other legal private entity. Eligible landowners must have an approved Forest Stewardship Plan and own 1,000 or fewer acres of qualifying land. Authorization may be obtained for exception of up to 5,000 acres.

Sale of Inventory Farmland

The Farm Service Agency advertises acquired farm property within 15 days of acquisition. Eligible beginning farmers and ranchers are given first priority to purchase these properties at the appraised market value for the first 75 days after acquisition.

Joint Financing Plan

Beginning farmer or rancher applicants may choose to participate in a joint financing plan that is also available to other applicants. In this program, the FSA lends up to 50% or more. The FSA may charge an interest rate of not less than 4%.

National United States Department Of Agriculture Offices

United States Department of Agriculture Information Center:
202-720-2791

Alternative Agricultural Research and Commercialization Center
0156 South Building, USDA, Ag Box 0401, Washington, DC 20250-0401
202-690-1655

Cooperative State Research, Education and Extension Service
USDA, Ag Box 2201, Washington, DC 20250-2201
202-720-4423

Animal and Health Inspection Service
USDA, USDA Center, Riverdale, MD 20737
301-734-8119

Packers and Stockyards Programs
Room 3039 South Bldg, USDA, Washington, DC 20250
202-720-7051

Forest Service
USDA, PO Box 96090, Washington, DC 20090-6090
202-205-1657

Risk Management Agency
Ag Box 0525, Washington, DC 20013
202-690-2803

Farm Service Agency
PO Box 2415, Stop 0506, Washington, DC 20013
202-720-5237

Natural Resources Service Agency
USDA, PO Box 2890, Washington, DC 20013
202-720-1873

Local United States Department Of Agriculture Offices

USDA Alabama State Headquarters

Farm Service Agency, PO Box 235013, Montgomery, Alabama 36123
(334)279-3543 Fax (334)279-3550

Natural Resources Conservation Service
Phone: 334-887-4581, Fax: 334-887-4551

State Cooperative Extension Service
Auburn University, Duncan Hall, Auburn, Alabama 36849
205-826-444

USDA Alaska State Headquarters

Farm Service Agency
800 W. Evergreen, Suite 216, Palmer, Alaska 99645
(907)-745-7982 Fax (907)745-7984

Natural Resources Conservation Service
Phone: 907-271-2424, Fax: 907-271-3951

State Cooperative Extension Service
University Of Alaska, PO Box 95151, Fairbanks, Alaska 99701
907-474-7246

USDA Arizona State Headquarters

Farm Service Agency
77 E. Thomas Road, Suite 240, Phoenix, Arizona 85012
(602)640-5200 Fax (602)640-5180

Natural Resources Conservation Service
Phone: 602-280-8808, Fax: 602-280-8809

State Cooperative Extension Service
University of Arizona
College of Agriculture, Forbes Bldg, Tucson, Arizona 85721
907-474-7246

USDA Arkansas State Headquarters

Farm Service Agency
700 W. Capitol Road, Room 5416, Little Rock, Arkansas 72201
(501)324-5220 Fax (501)324-5895

Natural Resources Conservation Service
Phone: 501-324-5445, Fax: 501-324-5648

State Cooperative Extension Service
University of Arkansas
1201 McAlmont, Little Rock, AR 72203
501-373-2500

USDA California State Headquarters

Farm Service Agency
1303 J. Street, #300, Sacramento, California 95814
(916)498-5311 Fax (916)498-5932

Natural Resources Conservation Service
Phone: 916-757-8215, Fax: 916-757-8379

State Cooperative Extension Service
University of California
300 Lakeside Drive, 6th Floor, Oakland, CA 94612
510-987-0060

USDA Colorado State Headquarters

Farm Service Agency
655 Parfet, Suite E305, 3rd Floor, Lakewood, Colorado 80215
(303)236-2866 Fax (303)236-2879

Natural Resources Conservation Service
Phone: 303-236-2886, Fax: 303-236-2896

State Cooperative Extension Service
Colorado State University, Fort Collins, CO 80523
303-491-6281

USDA Connecticut State Headquarters

Farm Service Agency
88 Day Hill Road, Windsor, Connecticut 06103
(860)285-8483 Fax (860)285-8481

Natural Resources Conservation Service
Phone: 860-487-4029, Fax: 860-487-4054

State Cooperative Extension Service
University of Connecticut
1376 Storrs Road, W.B. Young, Box U-36, Storrs, CT 06268
203-486-4125

USDA Delaware Headquarters

Farm Service Agency
1201 College Park Drive, #101,Dover, Delaware 19904
(302)678-2547 Fax (916)498-5932

Natural Resources Conservation Service
Phone: 302-678-4160, Fax: 302-678-0843

State Cooperative Extension Service
University of Delaware
Townsend Hall, Newark, DE 19717
302-451-2504

USDA Florida State Headquarters

Farm Service Agency
7770 NW 25th Street, Suite 1, Gainesville, Florida
(352)379-4500 Fax (352)379-4580

Natural Resources Conservation Service
Phone: 352-338-9500, Fax: 352-338-9574

State Cooperative Extension Service
University of Florida
Institute of Food and Agricultural Sciences, McCarty Hall, Gainesville, FL 32611
904-392-1761

USDA Georgia State Headquarters

Farm Service Agency
PO Box 1907, Athens, Georgia 30603
(706)546-2266 Fax (706)546-2014

Natural Resources Conservation Service
Phone: 706-546-2272, Fax: 706-546-2145

State Cooperative Extension Service
University of Georgia, College of Agriculture, Athens, GA 30602
404-542-3824

USDA Hawaii State Headquarters

Farm Service Agency
PO Box 50008, Honolulu, Hawaii 96850
(808)541-2644 Fax (808)541-2648

Natural Resources Conservation Service
Phone: 808-541-2600, Fax: 808-541-1335

State Cooperative Extension Service
University of Hawaii at Manoa, Honolulu, HI 96822
808-948-8397

USDA Idaho State Headquarters

Farm Service Agency
3220 Elder Street, Boise, Idaho 83705
(208)378-5650 Fax (208)378-5678

Natural Resources Conservation Service
Phone: 208-334-1601, Fax: 208-378-5735

State Cooperative Extension Service
University of Idaho
Agricultural Science Bldg., Moscow, ID 83843
208-885-6639

USDA Illinois State Headquarters

Farm Service Agency
3500 Wabash Ave., Springfield, Illinois 62707
(217)492-4180 Fax (217)492-4508

Natural Resources Conservation Service
Phone: 217-398-5310, Fax: 217-398-5310

State Cooperative Extension Service
University Of Illinois
1301 W. Gregory, Mumford Hall, Urbana, IL 61801
217-333-2660

USDA Indiana State Headquarters

Farm Service Agency
5981 Lakeside Blvd., Indianapolis, Indiana 46204
(317)290-3030 Fax (317)290-3030

Natural Resources Conservation Service
Phone: 317-290-3200, Fax: 317-290-3225

State Cooperative Extension Service
Purdue University
West Lafayete, IN 47906
317-494-8488

USDA Iowa State Headquarters

Farm Service Agency
10500 Buena Vista Court, Des Moines, Iowa 50322
(515)254-1540 Fax (515)254-1573

Natural Resources Conservation Service
Phone: 515-284-6655, Fax: 515-284-4394

State Cooperative Extension Service
Iowa State University
110 Curtis Hall, Ames, IA 50011
515-294-4576

USDA Kansas State Headquarters

Farm Service Agency
3600 Anderson Ave. , Manhattan, Kansas 66503
(913)539-3531 Fax (913)537-9659

Natural Resources Conservation Service
Phone: 913-823-4565, Fax: 913-823-4540

State Cooperative Extension Service
Kansas State University
Umberger Hall, Manhattan, KS 66506
913-532-5820

USDA Kentucky State Headquarters

Farm Service Agency
771 Corporate Drive, Suite 1000, Lexington, Kentucky 40503
(606)224-7601 Fax (606)224-7691

Natural Resources Conservation Service
Phone: 606-224-7350, Fax: 606-224-7399

State Cooperative Extension Service
University of Kentucky
College of Agriculture, Lexington, KY 40506
606-257-9000

USDA Louisiana State Headquarters

Farm Service Agency
3737 Government St., Alexandria, Louisiana 71302
(318)473-7650 Fax (318)473-7735

Natural Resources Conservation Service
Phone: 318-473-7751, Fax: 318-473-7682

State Cooperative Extension Service
Louisiana State University
Knapp Hall, University Station, Baton Rouge, LA 70803
504-388-6083

USDA Maine State Headquarters

Farm Service Agency
PO Box 406, Bangor, Maine 04402
(207)990-9140 Fax (207)990-9169

Natural Resources Conservation Service
Phone: 207-866-7241, Fax: 207-866-7262

State Cooperative Extension Service
University of Maine, Winslow Hall, Orono, ME 04169
207-581-3191

USDA Maryland State Headquarters

Farm Service Agency
River Center - 8335 Guilisford Road, Suite E, Columbia, Maryland 21046
(401)381-4550 Fax (401)962-4860

Natural Resources Conservation Service
Phone: 410-757-0861, Fax: 410-757-0687

State Cooperative Extension Service
University of Maryland
College Park, MD 20742
301-454-3722

USDA Massachusetts State Headquarters

Farm Service Agency
100 Cambridge Street, Boston, Massachusetts 01002
(413)256-0232 Fax (413)256-6290

Natural Resources Conservation Service
Phone: 413-253-4351, Fax: 413-253-4375

State Cooperative Extension Service
University of Massachusetts
College of Food and Natural Resources, Amherst, MA 01002
413-545-4500

USDA Michigan State Headquarters

Farm Service Agency
611 W. Ottawa, PO Box 30017, Lansing, Michigan 48909
(517)335-3401 Fax (517)335-1423

Natural Resources Conservation Service
Phone: 517-337-6701, Fax: 517-337-6905

State Cooperative Extension Service
Michigan State University
Agriculture Hall, East Lansing, MI 48824
517-355-2308

USDA Minnesota State Headquarters

Farm Service Agency
400 Agri Bank Building - 375 Jackson Street, St. Paul, Minnesota
(612)290-3660 Fax (612)290-3186

Natural Resources Conservation Service
Phone: 612-602-7856, Fax: 612-602-7914

State Cooperative Extension Service
University Of Minnesota
Coffey Hall, St. Paul, MN 55108
612-373-1223

USDA Mississippi State Headquarters

Farm Service Agency
PO Box 14995, Jackson, Mississippi 39236
(601)965-4966 Fax (601)965-4790

Natural Resources Conservation Service
Phone: 601-965-5205, Fax: 601-965-5178

State Cooperative Extension Service
Mississippi State University
Box 9601, Mississippi State, MS 39762
601-325-3036

USDA Missouri State Headquarters

Farm Service Agency
601 Business Loop 70 West, #225, Columbia, Missouri 65203
(573)876-0925 Fax (573)876-0935

Natural Resources Conservation Service
Phone: 573-876-0901, Fax: 573-876-0913

State Cooperative Extension Service
University of Missouri
University Hall, Columbia, MO 65211
314-751-3359

USDA Montana State Headquarters

Farm Service Agency
PO Box 670, Bozeman, Montana 59771
(406)587-6879 Fax (406)587-6877

Natural Resources Conservation Service
Phone: 406-587-6811, Fax: 406-587-6761

State Cooperative Extension Service
Montana State University
College of Agriculture, Bozeman, MT 59717
406-994-3681

USDA Nebraska State Headquarters

Farm Service Agency
PO Box 57975, Lincoln, Nebraska 68505
(402)437-5284 Fax (402)437-5280

Natural Resources Conservation Service
Phone: 402-437-5300, Fax: 402-437-5327

State Cooperative Extension Service
University of Nebraska
Agriculture Hall, Lincoln, NB 68583
402-472-2966

USDA Nevada State Headquarters

Farm Service Agency
1755 E. Plumb Lane, #202, Reno, Nevada 89502
(702)784-5374 Fax (702)784-5376

Natural Resources Conservation Service
Phone: 702-784-5863, Fax: 702-784-5939

State Cooperative Extension Service
University Of Nevada
College Of Agriculture, Reno, Nevada 89557

USDA New Hampshire State Headquarters

Farm Service Agency
22 Bridge Street - 4th Floor, Concord, New Hampshire 03301
(603)224-7941 Fax (603)225-1410

Natural Resources Conservation Service
Phone: 603-868-7581, Fax: 603-868-5301

State Cooperative Extension Service
University of New Hampshire
College of Life Sciences and Agriculture, Durnham, NH 03824
603-862-1520

USDA New Jersey State Headquarters

Farm Service Agency - New Jersey State Headquarters
163 Route 130, Building 2, Suite E, Bordentown, New Jersey 08505
(602)298-3446 Fax (602)298-8780

Natural Resources Conservation Service
Phone: 908-246-1662, Fax: 908-246-2358

State Cooperative Extension Service
Rutgers University
Cook College, Box 231, New Brunswick, NJ 08903
201-932-9306

USDA New Mexico State Headquarters

Farm Service Agency
6200 Jefferson Street NE, Albuquerque, New Mexico 87109
(505)646-2804 Fax (505)761-4934

Natural Resources Conservation Service
Phone: 505-766-2173, Fax: 505-761-4463

State Cooperative Extension Service
New Mexico State University
Box 3 AE, Las Cruces, NM 88003
505-646-1806

USDA New York State Headquarters

Farm Service Agency
441 W. Salina Street, Suite 356, Syracuse, New York 13202
(315)477-6303 Fax (315)477-6323

Natural Resources Conservation Service
Phone: 315-477-6504, Fax: 315-477-6550

State Cooperative Extension Service
Cornell University
Roberts Hall, Ithaca, NY 14853
607-255-2117

USDA North Carolina State Headquarters

Farm Service Agency
4407 Bland Road, Suite 175, Raleigh, North Carolina 27609
(919)790-2957 Fax (919)790-2954

Natural Resources Conservation Service
Phone: 919-873-2101, Fax: 919-873-2155

State Cooperative Extension Service
North Carolina State University
112 Patterson Hall, Box 7601, Raleigh, NC 27695
919-515-2668

USDA North Dakota State Headquarters

Farm Service Agency
PO Box 3046, Fargo, North Dakota 58108
(701)239-5224 Fax (701)239-5696

Natural Resources Conservation Service
Phone: 701-250-4421, Fax: 701-250-4778

State Cooperative Extension Service
North Dakota State University
Box 5437, Fargo, ND 58105
701-237-8944

USDA Ohio State Headquarters

Farm Service Agency
200 N. High, Room 540, Columbus, Ohio 43215
(401)381-4550 Fax (401)962-4860

Natural Resources Conservation Service
Phone: 614-469-6962, Fax: 614-469-2083

State Cooperative Extension Service
Ohio State University
2120 Fyffe Road, Columbus, OH 43210
614-292-4067

USDA Oklahoma State Headquarters

Farm Service Agency
100 USDA, Suite 102, Stillwater, Oklahoma 74074
(405)742-1130 Fax (405)742-1177

Natural Resources Conservation Service
Phone: 405-742-1204, Fax: 405-742-1201

State Cooperative Extension Service
Oklahoma State University, Agriculture Hall, Stillwater, OK 74078
415-624-5400

USDA Oregon State Headquarters

Farm Service Agency
PO Box 1300, Tualatin, Oregon 97062
(503)692-6830 Fax (503)692-8139

Natural Resources Conservation Service
Phone: 503-414-3200, Fax: 503-414-3277

State Cooperative Extension Service
Oregon State University
Balland Hall, Corvallis, Oregon 97331
503-754-2711

USDA Pennsylvania State Headquarters

Farm Service Agency
1 Credit Union Plaza, Suite 320, Harrisburg, Pennsylvania 17110
(717)782-4457 Fax (717)782-4813

Natural Resources Conservation Service
Phone: 717-782-3445, Fax: 717-782-4469

State Cooperative Extension Service
Penn State University
323 Agriculture Bldg, University Park, PA 16802
814-863-3438

USDA Puerto Rico State Headquarters

Farm Service Agency
PO Box 11188, Fernandez Junzos Station, Santurce, Puerto Rico 00910
(809)729-6902 Fax (809)729-6827

Natural Resources Conservation Service
Phone: 809-766-5206, Fax: 809-766-5987

State Cooperative Extension Service
University of Puerto Rico, Rio Piedras, PR 00928

USDA Rhode Island State Headquarters

Farm Service Agency
60 Quaker Lane, West Bay Office Complex, Room 40, Warwick, Rhode Island 02886
(401)828-8232 Fax (401)528-5206

Natural Resources Conservation Service
Phone: 401-828-1300, Fax: 401-828-0433

State Cooperative Extension Service
University of Rhode Island
Woodward Hall, Kingston, RI 02881
401-792-2815

USDA South Carolina State Headquarters

Farm Service Agency
1927 Thurmond Mall, Suite 100, Columbia, South Carolina 29201
(803)806-3830 Fax (803)806-3839

Natural Resources Conservation Service
Phone: 803-253-3935, Fax: 803-253-3670

State Cooperative Extension Service
Clemson University
Barre Hall, Clemson, SC 29634
803-656-3382

USDA South Dakota State Headquarters

Farm Service Agency
200 4th Street SW, Federal Building, Room 308, Huron, South Dakota 57350
(605)352-1169 Fax (605)352-1195

Natural Resources Conservation Service
Phone: 605-353-1783, Fax: 605-352-1270

State Cooperative Extension Service
South Dakota State University
Box 2207, Brookings, SD 57007
605-688-4792

USDA Tennessee State Headquarters

Farm Service Agency
801 Broadway, US Courthouse, Room 579, Nashville, Tennessee 37204
(615)736-5121 Fax (615)736-7801

Natural Resources Conservation Service
Phone: 615-736-5471, Fax: 615-736-7764

State Cooperative Extension Service
University of Tennessee
Box 1071, Morgan Hall, Knoxville, TN 37901
615-974-7114

USDA Texas State Headquarters

Farm Service Agency
PO Box 2900, College Station, Texas 77840
(409)260-9424 Fax (409)260-9358

Natural Resources Conservation Service
Phone: 254-298-1214, Fax: 254-298-1388

State Cooperative Extension Service
Utah State University
Logan, UT 84322
801-750-2200

USDA Utah State Headquarters

Farm Service Agency
125 S. State Street, Room 4239, Salt Lake City, Utah 84138
(801)524-5013 Fax (801)524-5244

Natural Resources Conservation Service
Phone: 801-524-5050, Fax: 801-524-4403

State Cooperative Extension Service
University of Vermont
College Of Agriculture, 178 S. Prospect St., Burlington, VT 05401
802-656-3036

USDA Vermont State Headquarters

Farm Service Agency
346 Shelburne, Executive Square Building, Burlington, Vermont 05401
(802)658-2803 Fax (802)660-0953

Natural Resources Conservation Service
Phone: 802-951-6795, Fax: 802-951-6327

State Cooperative Extension Service
University of Vermont
College of Agriculture, 178 S. Prospect St., Burlington, VT 05401
802-656-3036

USDA Virginia State Headquarters

Farm Service Agency
1606 Santa Rosa Road, Culpepper Building, Suite 138, Richmond, Virginia 23229
(804)287-1522 Fax (804)287-1723

Natural Resources Conservation Service
Phone: 804-287-1671, Fax: 804-287-1737

State Cooperative Extension Service
Virginia Polytechnic Institute
Burruss Hall, Blacksburg, VA 24061
703-961-6705

USDA Washington State Headquarters

Farm Service Agency
W 316 Boone, Suite 568, Rockpointe Tower, Spokane, Washington 99201
(306)902-1800 Fax (306)902-2092

Natural Resources Conservation Service
Phone: 509-323-2900, Fax: 509-323-2909

State Cooperative Extension Service
Washington State University
College of Agriculture, Pullman, WA 99164
509-335-2811

USDA West Virginia State Headquarters

Farm Service Agency
PO Box 1049, Morganville, West Virginia 25305
(304)558-2201 Fax (304)558-0451

Natural Resources Conservation Service .
Phone: 304-291-4153, Fax: 304-291-4628

State Cooperative Extension Service
West Virginia University
Morgantown, WV 26506
304-293-5691

USDA Wisconsin State Headquarters

Farm Service Agency
6515 Watts Road, Room 100, Madison, Wisconsin 53719
(608)276-8732 Fax (608)271-9425

Natural Resources Conservation Service
Phone: 608-264-5341, Fax: 608-264-5483

State Cooperative Extension Service
University of Wisconsin
423 N. Lake St., Madison, WI 53706
608-263-2775

USDA Wyoming State Headquarters

Farm Service Agency
951 Werner Court, Suite 130, Casper, Wyoming 82002
(307)261-5230 Fax (307)261-5857

Natural Resources Conservation Service
Phone: 307-261-6453, Fax: 307-261-6490

University of Wyoming
Box 3354, University Station
Laramie, WY 82071
307-766-5124

Chapter 9 – Federal Help For Rural Living

The USDA's Rural Development programs were created to improve the economy and quality of life for rural America. To spur growth in rural areas, both private and business and to improve rural housing and employment conditions. It is a misconception on the part of the vast majority of the public that poverty is an inner city problem. When in fact the majority of the nation's poor are located in rural areas. More than 53 million people live in rural America nearly 16% earn wages below the Federal poverty level. There are 2.5 million substandard housing units, compared to 2.4 million in cities and 1.2 million in the suburbs. More than 418,000 households still lack running water.

Home Ownership Loans: To aid low income and moderate income rural residents to purchase, construct, repair or relocate a dwelling and related facilities. New no down payment rural residential loans available.

Rural Rental Housing Loans: To allow individuals or organizations to build or rehabilitate rental units for low-income and moderate-income residents in rural areas.

Rental Assistance: To reduce out-of-pocket cash that low-income families pay for rent, including utilities.

Home Improvement and Repair Loans and Grants: To enable low income rural homeowners to remove health and safety hazards in their homes and to make homes accessible for people with disabilities. Grants are available for people over 62 years old and older who cannot afford to repay a loan.

Self-Help Housing Loans: To assist groups of six to eight low income families in helping each other build homes by providing materials, site, and the skilled labor they cannot furnish. The families must agree to work together until all the homes are finished.

Rural Housing Site Loans: To buy adequate building sites for development into a desirable community by private or public nonprofit organization.

Housing Preservation Grants: To provide qualified public nonprofit organizations and public agencies with grant funds for effective programs to assist very low and low income homeowners repair and rehabilitate their homes in rural areas and to assist rental property owners and co-ops repair and rehabilitate their units if they agree to make such units available to low and very low income persons.

Business and Industrial Loans: Provides direct and guaranteed loans designed to create and save rural jobs and improve the economic and environmental climates of rural communities under 50,000 population. Eligible borrowers include private for profit and nonprofit organizations.

VERY LOW TO MODERATE INCOME HOUSING LOANS (Section 502 Rural Housing Loans)

Agency: Rural Housing Service, USDA

Range & Average of Financial Assistance: $1,000 to $105,000, an average of $59,106 for new construction, and $56,249 for existing cost. Guaranteed loans in high cost areas may be higher.

Objective: To assist lower-income rural families through direct loans to buy, build, rehabilitate, or improve and to provide the applicant with modest, decent, safe, and sanitary dwellings and related facilities as a permanent residence. Subsidized funds are available only on direct loans for low and very low-income applicants. Nonsubsidized funds (loan making) are available for very low- and low-income applicants who are otherwise eligible for subsidy, but at the present time, the subsidy is not needed. Loan guarantees are also available to assist low and moderate income rural families in home acquisition.

Type of Assistance: Direct Loans, Guaranteed/Insured Loans

Summary: Direct loans may be used for construction, repair or purchase of housing; to provide adequate sewage disposal facilities and/or safe water supply for the applicant's household; for weatherization; to purchase or install essential equipment if the equipment is normally sold with dwellings in the area; to buy a minimum adequate site on which to place a dwelling for the applicant's own use; and under certain conditions to finance a manufactured home and its site. Housing debts may under certain circumstances be refinanced with direct loans.
Dwellings financed must be modest, decent safe and provide sanitary housing. Cost of dwelling financed cannot exceed the maximum dollar limitation established under section 2036(b) of the National Housing Act for the area in which the property is located. The property must be located in a place that is rural in character and does not exceed 10,000 population or in certain cases a place whose population exceeds 10,000 but is not in excess of 25,000, provided the place has a serious lack of mortgage credit for low- and moderate-income families as determined by the Secretary of Agriculture and the Secretary of Housing and Urban Development. Assistance is available in the States, the Commonwealth of Puerto Rico, the U.S. Virgin Islands, Guam, American Samoa, the Commonwealth of Northern Marinas, and the Trust Territories of the Pacific Islands. Loans are made at the interest rate(s) specified in FmHA Instruction 440.1, Exhibit B (available in any FmHA county office) for the type of assistance involved, and are repaid over an amortization period of up to 33 years for regular loans and 38 years for loans to applicants whose adjusted annual income do not exceed 60% of the area median income, if necessary to show repayment ability. Payment assistance is granted annually which would reduce the monthly installment on the note to an amount equal to what it would be if the note were amortized to as low

as 1%, depending on the loan amount, the size and income of the family. Payment assistance is subject to recapture by the government upon liquidation of the account. The Deferred Mortgage Demonstration Program was available during fiscal years 1991-1995 however, there was no funding provided for deferred mortgage authority or loans for deferred mortgage assumptions. Deferred mortgage payment assistance is available to make home ownership affordable for a greater number of very low-income families who lack repayment for the mortgage when amortized at 1% for a 38 year period, or 30 years for a manufactured home. The guaranteed program is an acquisition only program. Guaranteed loans are amortized over 30 years. The interest rate is negotiated with the lender.

Application Process: For direct loans, applicants must file loan applications at the RD county office serving the county where the dwelling is or will be located. For guaranteed loans, applicants must contact a local lender.

VERY LOW INCOME HOUSING REPAIR LOANS AND GRANTS (SECTION 504 RURAL HOUSING LOANS AND GRANTS)

Agency: Rural Housing Service, USDA

Range & Average of Financial Assistance: Loans to $4,821, Grants to $3,995

Objective: To give low-income rural homeowners an opportunity to make essential repairs to their homes to make them safe and to remove health hazards to the family or the community.

Type of Assistance: Direct Loans, Project Grants

Summary: To assist very low-income owner-occupants in rural areas to repair or improve their dwellings. Grant funds may only be used to make such dwellings safe and sanitary and to remove health and safety hazards. This includes repairs to the foundation, roof or basic structure as well as water and waste disposal systems, and weatherization. Loans bear an interest rate of 1%
and are repaid over a period of up to 20 years. In addition to the above purpose, loan funds may be used to modernize the dwelling. Maximum loan amount cannot exceed a cumulative total of $20,000 to any eligible person and maximum lifetime grant assistance is $7,500 to any eligible person 62
years of age or older for home improvement. The house must be located in a place which is rural in character and does not exceed 10,000 population. Some places with population between 10,000 and 25,000 may be eligible if not within a Metropolitan Statistical Area (MSA) and has a serious lack of mortgage credit for low and moderate-income families as determined by the Secretary of
Agriculture and the Secretary of Housing and Urban Development. Assistance is available in States, Puerto Rico, the U.S. Virgin Islands, Guam, American Samoa, the Commonwealth of Northern Marianas, and the Trust Territories of
the Pacific Islands. Applicants must own and occupy a home in a rural area; and be a citizen of the United States or reside in the United States after having been legally admitted for permanent residence or on indefinite parole. Loan recipients must have sufficient income to repay the loan. Grant recipients must be 62 years of age or older and be unable to repay a loan for that part of the
assistance received as a grant. Applicant's income may not exceed the very low-income limit set forth in FmHA Instructions. Very low-income limits range from $6,300 to $22,650 for a single person household, depending on an area's median income.

Application Process: Applicants must file Form RD 410-4 at the RHS/RD county office serving the county where the dwelling is located.

RURAL RENTAL ASSISTANCE PAYMENTS (RENTAL ASSISTANCE)

Agency: Rural Housing Service, USDA

Range & Average of Financial Assistance: No data available.

Objective: To reduce the tenant contribution paid by low-income families occupying eligible Rural Rental Housing (RRH), Rural Cooperative Housing (RCH), and Farm Labor Housing (LH) projects financed by the Rural Housing Service (RHS) through its Sections 515, 514 and 516 loans and grants.

Type of Assistance: Direct Payments for Specified Use.

Summary:Rental assistance may be used to reduce the rents paid by low-income senior citizens or families and domestic farm laborers and families whose rents exceed 30% of an adjusted annual income which does not exceed the limit established for the State as indicated in 7 CFR Exhibit C to Part 1944, Subpart A (FmHA Instruction 1944-A, Exhibit C). Tenants who may be eligible must occupy units in eligible RRH, RCH and LH projects financed by RNS.

Application Process: This program is eligible for coverage under E.O. 12372, "Intergovernmental Review of Federal Programs." An applicant should consult the office or official designated as the single point of contact in his or her State for more information on the process the State requires to be followed in applying for assistance, if the State has selected the program for review. This program is excluded from coverage under 7 CFR 3015 or 3016. Eligible borrowers will initiate the processing by submitting Form FmHA 1944-25, "Request for Rental Assistance". Applications shall then be reviewed under the procedure set forth in Exhibit E of FmHA Instruction 1930-C.

RURAL HOUSING SITE LOANS AND SELF HELP HOUSING LAND DEVELOPMENT LOANS (Section 523 and 524 Site Loans)

Agency: Rural Housing Service, USDA

Range & Average of Financial Assistance: (523) $9,380, (524) $12,000

Objective: To assist public or private nonprofit organizations interested in providing sites for housing, to acquire and develop land in rural areas to be subdivided as adequate building sites and sold on a cost development basis to families eligible for low and very low income loans, cooperatives, and broadly based nonprofit rural rental housing applicants.

Type of Assistance: Direct Loans

Summary:For the purchase and development of adequate sites, including necessary equipment which becomes a permanent part of the development; for water and sewer facilities if not avail- able; payment of necessary engineering, legal fees, and closing costs; for needed landscaping and other necessary facilities related to buildings such as walks, parking areas, and driveways. Restrictions: loan limitation of $200,000 without national office approval; loan funds may not be used for refinancing of debts,

payment of any fee, or
commission to any broker, negotiator, or other person for the referral of a prospective applicant or solicitation of a loan; no loan funds will be used to pay operating costs or expenses of administration other than actual cash cost of incidental administrative expenses if funds to pay those expenses are not otherwise available. Repayment of loan is expected within two years.

Application Process: Environmental impact assessment and environmental impact statements are required for this program. This program is eligible for coverage under E.O. 12372, "Intergovernmental Review of Federal Programs." An applicant should consult the office or official designated as the single point of contact in his or her State for more information on the process the State requires to be followed in applying for assistance, if the State has selected the program for review. The application will be in the form of a letter to the Rural Development Manager of the Rural Development (RD). Supporting information and costs should be included as needed.

RURAL RENTAL HOUSING LOANS

Agency: Rural Housing Service, USDA

Range & Average of Financial Assistance: Initial insured loans to individuals, $60,000 to $450,000; $250,000. Initial insured loans to organizations, $75,000 to $2,000,000; $950,000.

Objective: To provide economically designed and constructed rental and cooperative housing and related facilities suited for rural residents.

Type of Assistance: Direct Loans

Summary:Loans can be used to construct, purchase and substantially rehabilitate rental or cooperative housing or to develop manufactured housing projects. Housing as a general rule will consist of multi-units with two or more family units and any appropriately related facilities. Funds may also be used to provide approved recreational and service facilities appropriate for use in
connection with the housing and to buy and improve the land on which the buildings are to be located. Loans may not be made for nursing, special care, or institutional-type homes. Applicants may be individuals, cooperatives, nonprofit organizations, State or local public agencies, profit corporations, trusts, partnerships, limited partnerships, and be unable to finance the housing either with their own resources or with credit obtained from private sources.
However, applicants must be able to assume the obligations of the loan, furnish adequate security, and have sufficient income for repayment. They must also have the ability and intention of maintaining and operating the housing for purposes for which the loan is made. Loans may be made in communities up to 10,000 people in MSA areas and some communities up to 20,000 population in
non-MSA areas. Applicants in towns of 10,000 to 20,000 should check with their local Rural Development; office to determine if the agency can serve them. Assistance is available to eligible applicants in States, Puerto Rico, the Virgin Islands, Guam, American Samoa, the Northern Marianas, and the Trust Territory of the Pacific Islands. Occupants must be very low-, low- or moderate-income families households, elderly, handicapped, or disabled persons.

Application Process: Applications are subject to an environmental impact assessment. An environmental impact statement is required for this program when there is a significant impact on the environment. This program is eligible for coverage under E.O. 12372, "Intergovernmental Review of Federal Programs." An applicant should consult the office or official designated as the single point of contact in his or her State for more information on the process the State requires to be followed in applying for assistance, if the State has selected the program for review. This program is excluded from coverage under 7 CFR 3016. The application will be on SF 424.2 "Application for Federal Assistance"
which may be submitted to the RD county office where the housing will be located but should be submitted to the office having jurisdiction. Appropriate attachments such as preliminary market data, cost estimates, and financial statement and plans, if available, should be included, as per FmHA
1944-E exhibit A.

RURAL SELF HELP HOUSING TECHNICAL ASSISTANCE (SECTION 523 TECHNICAL ASSISTANCE)

Agency: Rural Housing Service, USDA

Range & Average of Financial Assistance: Fiscal year 1996 to $203,692

Objective: To provide financial support for programs of technical and supervisory assistance that will aid needy very low and low-income individuals and their families in carrying out mutual self-help housing efforts in rural areas.

Type of Assistance: Project Grants

Summary: Not-for-profit organizations may use technical assistance funds to hire the personnel to carry out a program of technical assistance for self-help housing in rural areas; to pay necessary and reasonable office and administrative expenses; to purchase or rent equipment such as power tools for use by families participating in self-help housing construction; and to pay
fees for training self-help group members in construction techniques or for other professional services needed. Funds will not be used to hire personnel to perform any construction work, to buy real estate or building materials, or pay any debts, expenses or costs other than previously outlined for participating families in self-help projects.

Application Process: The standard application forms as furnished by the Federal agency and required by Departmental Regulations 3015 and 3016 must be used for this program. This program is eligible for coverage under E.O. 12372, "Intergovernmental Review of Federal Programs." An applicant should consult the office or official designated as the single point of contact
in his or her State for more information on the process the State requires to be followed in applying for assistance, if the State has selected the program for review. Form SF-424 is submitted to the District office of the Rural Development
(RD), including information attached to Part IV, fully documenting the applicant's experience, need for the grant and the proposed amount needed. Attachments relating to the size of the proposed project, estimated cost, budget and need are also required.

RURAL HOUSING PRESERVATION GRANTS

Agency: Rural Housing Service, USDA

Range & Average of Financial Assistance: For fiscal year 95, 207 preapplications were funded, to assist 4,868 units.

Objectives: To assist very low- and low-income rural residents individual homeowners, rental property owners (single/multi-unit) or by providing the consumer cooperative housing projects (co-ops) the necessary assistance to repair or rehabilitate their dwellings. These objectives will be accomplished through the establishment of repair/rehabilitation, projects run by eligible applicants. This program is intended to make use of and leverage any other available housing programs which provide resources to very low and low-income rural residents to bring their dwellings up to development standards.

Type of Assistance: Project Grants

Summary: Organizations may use less than 20% of the Housing Preservation Grant funds for program administration purposes, such as to hire the personnel to carry out a project of housing rehabilitation to meet the needs of very low and low-income persons in rural areas; to pay necessary and reasonable office and administrative expenses; and to pay reasonable fees for training of organization personnel. Eighty percent or more of funds must be used for loans, grants or other assistance on individual homes, homeowners, rental properties or co-ops to pay any part of the cost for repair or rehabilitation of structures; funds may not be used to hire personnel to perform construction or to pay any debts, expenses or costs other than previously outlined and approved in the project application.

Application Process: The standard application forms as furnished by the Federal agency and required by 7 CFR parts 3015 or 3016 must be used for this program. Preapplications on SF 424.1 "Application for Federal Assistance (for non-construction)," must be submitted to RD. Applicants are encouraged to consult with the RD District or State office prior to submission of a preapplication and to receive assistance in the preparation of their preapplication. An environmental impact assessment is required for this program. This program is eligible for coverage under E.O. 12372, "Intergovernmental Review of Federal Programs." An applicant should consult the office or official designated as the single point of contact in his or her State for more information on the process the State requires to be followed in applying for assistance, if the State has selected the program for review. Applicants must file a preapplication form. The standard application forms as furnished by the Agency and required by 7 CFR parts 3015 or 3016 must be used for this program. Upon notification by Form AD-622, "Notice of Preapplication Review Action," that the applicant has been tentatively selected for funding under the preapplication project selection criteria,
the applicant may submit an application on SF 424.1, "Application For Federal Assistance (for non-construction)" to the RD District office.

SECTION 538 RURAL RENTAL HOUSING GUARANTEED LOANS

Agency: Rural Housing Service, USDA

Range & Average of Financial Assistance: NA

Objective: This program has been designed to increase the supply of affordable multifamily housing through partnerships between RHS and major lending sources, as well as State and local housing finance agencies and bond insurers. The program provides effective new forms of Federal credit enhancement for the development of affordable multifamily housing by lenders.

Type of Assistance: Guaranteed/Insured Loans

Summary: The guarantee will encourage the construction of new rural rental housing and appropriate related facilities. Housing as a general rule will consist of multi-units with two or more family units. The guarantee may not be made for nursing, special care or industrial type housing. The applicant in this program is the lender that will use the guarantee as a credit enhancement and therefore be more likely to make the loan. The lender must be approved by Fannie Mae, Freddie Mac, HUD or be a State Housing Finance Agency. The projects must be located in rural areas as defined by the Agency. Beneficiary Eligibility: Occupants must be very-low, low- or moderate-income households, elderly, handicapped, or disabled persons with income not in excess of 115% of the Median Income.

Application Process: The lender originates the loan and performs the necessary underwriting and provides the documentation required by the RHS and request for guarantee to the RHS for consideration.

BUSINESS AND INDUSTRIAL LOANS

Agency: Rural Business Cooperative Service, USDA

Range & Average of Financial Assistance: $35,000 to $10,000,000, $1,140,000 (average size) for B&I guaranteed loans, $500,000 (average size) est for B&I direct loans.

Objective: To assist public, private, or cooperative organizations (profit or nonprofit), Indian tribes or individuals in rural areas to obtain quality loans for the purpose of improving, developing or financing business, industry, and employment and improving the economic and environmental climate in rural communities including pollution abatement and control.

Type of Assistance: Direct Loans, Guaranteed/Insured Loans

Summary: Financial assistance may be extended for: (a) Business and industrial acquisition, construction, conversion, enlargement, repair, modernization, development costs; (b) purchasing and development of land, easements, rights-of-way, buildings, facilities, leases or materials; (c) purchasing equipment, leasehold/improvements, machinery and supplies; and (d) pollution control and abatement. Maximum loan size is $10,000,000 and maximum time allowable for final maturity is limited to 30 years for land and buildings, the usable life of machinery and equipment purchased with loan funds, not to exceed 15 years, and 7 years for working capital. Interest rates on guaranteed loans are negotiated between the lender and the borrower. For loans of $2 million or less, the maximum percentage of guarantee is 90%. For loans over $2 million but not over $5 million, the maximum percentage of guarantee is 80%. For loans in excess of $5 million, the maximum percentage of guarantee is 70%. Losses on principal advanced, including protective advances, and accrued interest, may be guaranteed to the lender. Loans may not be made or guaranteed (a) to pay off a creditor in excess of the value of the collateral, (b)

for distribution or payment to the owner, partners, shareholders, or beneficiaries of the applicant or members of their families when such persons shall retain any portion of their equity in the business, (c) for projects involving agricultural production, (d) for transfer of ownership of a business unless the
loan will keep the business from closing, or prevent the loss of employment opportunities in the area, or provide expanded job opportunities, (e) for the guarantee of lease payments, (f) for financing community antenna television services or facilities, (g) for charitable and educational institutions, churches,
fraternal organizations, hotels, motels, tourist homes, convention centers, tourist, recreation or amusement facilities, lending and investment institutions and insurance companies, (h) for any legitimate business activity where more than 10% of the annual gross income is derived from legalized gambling, for guarantee of loans made by other Federal agencies except those made by
Banks for Co-ops, Federal Land Bank or Production Credit Associations, and (j) for any project which is likely to result in transfer of business or employment from one area to another or cause production which exceeds demand. Interested parties should contact the Rural Business Cooperative Service (RBS) or the Rural Development RD State Office nearest them. RD administers the program at the local level.

Application Process: This program is eligible for coverage under E.O. 12372, "Intergovernmental Review of Federal Programs." An applicant should consult the office or official designated as the single point of contact in his or her State for more information on the process the State requires to be followed in applying for assistance, if the State has selected the program for review. All preapplication letters must be coordinated fully with appropriate State agencies in keeping with Executive Order 12372, "Intergovernmental Review of Federal Programs," in a manner that will assure maximum support of the State's strategies for development of its rural areas. The application form as furnished by the Federal agency must be used for this program. An environmental assessment is required and an environmental impact statement may be required for this program. Form FmHA 449-1 is used for guaranteed loans and filed at the RD State office. This program is excluded from coverage under OMB Circular No. A-110.

National Headquarters

Rural Housing Service
USDA, Washington DC, 20250
202-720-1660

Rural Business-Cooperative Service
USDA, Washington, DC 20250-0700
(202) 690-4730.

Local Rural Development State Offices

Alabama Rural Development State Office,
4121 Carmichael Road, Sterling Center, Suite 601, Montgomery, Al 36106
Ph: 334-279-3400, Fax: 334-279-3484

Alaska Rural Development State Office
800 W. Evergreen, Suite 201, Palmer, AK 99645
Phone: 907-745-2176, Fax: 907-745-5398

Arkansas Rural Development State Office
USDA Service Center, Federal Building, Room 3416, 700 Capitol Avenue
Little Rock, AR 72201
Phone: 501-324-6281, Fax: 501-324-6346

Arizona Rural Development State Office
3003 N. Central Avenue Suite 900, Phoenix, AZ 85012
Phone: 601-280-8700, Fax: 601-280-8770

California Rural Development State Office
194 W. Main St, Suite , Woodland, CA 95695
Phone: 916-668-2000

Colorado Rural Development State Office
655 Parfet Street Room E 100, Lakewood, CO 80215
Phone: 303-236-2801, Fax: 303-236-2854

Delaware Rural Development State Office
4611 So. Dupont Hwy PO Box 400, Camden, DE 19934-9998
Phone: 302-697-4300, Fax: 302-697-4390

Florida Rural Development State Office
4440 N.W. 25th Place, Gainesville, FL 32614-7010
Phone: 904-338-3400, Fax: 904-338-3405

Georgia Rural Development State Office
355 E. Hancock Ave., Stephens Federal Building, Athens, GA 30610
Phone: 706-546-2173, Fax: 706-456-2162

Hawaii Rural Development State Office
154 Wainuenue Avenue, Federal Building, Rm 311, Hilo, HI 89701
Phone: 808-933-3000, Fax: 808-935-1590

Iowa Rural Development State Office
210 Walnut Street, Federal Building, Suite 873, Des Moines, IA 50309
Phone: 515-284-4663, Fax: 515-284-4859

Idaho Rural Development State Office
3232 Elder Street, Boise, ID 83705
Phone: 208-378-5600, Fax: 208-378-5643

Illinois Rural Development State Office
Illinois Plaza, Suite 103, 1817 S. Neil St., Champaign, IL 61820
Phone: 217-398-5237, Fax: 217-398-5337

Indiana Rural Development State Office
5975 Lakeside Blvd., Indianapolis, IN 46278
Phone: 317-290-3100, Fax: 317-290-3095

Kansas Rural Development State Office
1200 SW Executive Drive, PO Box 4653, Topeka, KS 66604
Phone: 913-271-2700, Fax: 913-271-2708

Kentucky Rural Development State Office
771 Corporate Plaza, Suite 200, Lexington, KY 40503
Phone: 606-224-7300, Fax: 606-224-7300

Louisiana Rural Development State Office
3727 Government Street, Alexandria, LA 71302
Phone: 318-473-7920, FAX: 318-473-7829

Maine Rural Development State Office
444 Stillwater Avenue, Suite 2, PO Box 405, Bangor, ME 04402-0405
Phone: 207-990-9160, Fax: 207-990-9165

Michigan Rural Development State Office
3001 Coolidge Road, Ste 200, E. Lansing, MI 48823
Phone: 517-337-6635, Fax: 517-337-6913

Missouri Rural Development State Office
601 Business Loop 70 West, Parkade Center, Suite 235, Columbia, MO 65203
Phone: 314-876-0976, Fax: 314-876-0977

Minnesota Rural Development State Office
410 Farm Credit Service Building, 375 Jackson Street, St. Paul, MN 55101-1853
Phone: 612-602-7800, Fax: 612-602-7824

Mississippi Rural Development State Office
Suite 831, Federal Building, 100W Capital SE, Jackson, MS 39269
Phone: 601-965-4318, Fax: 601-965-5384

Maryland Rural Development State Office
4611 So. Dupont Hwy, PO Box 400, Camden, DE 19934-9998
Phone: 302-697-4300, Fax: 302-697-4390

Massachusetts Rural Development State Office
451 West Street, Amherst, MA 01002
Phone: 413-253-4302, Fax: 413-253-4347

Montana Rural Development State Office
900 Technology Blvd, Suite B PO Box 850, Bozeman, MT 59771
Phone: 406-585-2580, Fax: 406-585-2565

North Carolina Rural Development State Office
4405 Bland Road, Raleigh, NC 27609
Phone: 919-873-2000, Fax: 919-873-2075

North Dakota Rural Development State Office
Federal Building, Room 208, 220 E. Rosser Ave., Bismark, ND 58502
Phone: 701-250-4781, Fax: 701-250-4670

Nebraska Rural Development State Office
Federal Building, Rm 308, 100 Centennial Mall N, Lincoln, NE 68508
Phone: 402-437-5551, Fax: 402-437-5408

New Jersey Rural Development State Office
Transfield Plaza, #22, Woodland Road Mt. Holly, NJ 08060
Phone: 609-265-3600, Fax: 609-265-3651

New Mexico Rural Development State Office
6200 Jefferson St, NE, Room 255, Albuquerque, NM 87109
Phone: 505-761-4950, Fax: 505-761-4976

Nevada Rural Development State Office
1390 S. Curry St., Carson, NV 89703
Phone: 702-887-1222, Fax: 702-885-0841

New York Rural Development State Office
The Galleries of Syracuse, 441 S. Salina St., 5th Floor, Syracuse, NY 13202
Phone: 315-477-6433

Ohio Rural Development State Office
Federal Building, Room 507, 200 North High St., Columbus, OH 43215
Phone: 614-469-5608, Fax: 614-469-5600

Oklahoma Rural Development State Office
100 USDA, Suite 108, Stillwater, OK 74074
Phone: 405-742-1000, Fax: 405-742-1005

Oregon Rural Development State Office
101 SW Main Street, Suite 1410, Portland, OR 97204
Phone: 503-414-3300, Fax: 503-414-3386

Pennsylvania Rural Development State Office
One Credit Union Place, Suite 330, Harrisburg, PA 17110-4476
Phone: 717-782-4476, Fax: 717-782-4883

Puerto Rico Rural Development State Office
159 Carlos E. Chardon St., Hato Rey, PR 00918-5481
Phone: 809-766-5095, Fax: 809-766-5844

Massachusetts Rural Development State Office
451 West Street, Amherst, MA 01022
Phone: 413-253-4302, Fax: 413-253-4347

South Carolina Rural Development State Office
Strom Thurmond Federal Building, 1835 Assembly St., Room 1007, Columbia, SC 29201
Phone: 803-253-3725, Fax: 803-765-5633

South Dakota Rural Development State Office
Federal Building, Room 308, 200 4th Street SW, Huron, SD 57350
Phone: 605-352-1100, Fax: 605-352-1146

Tennessee Rural Development State Office
3322 West End Ave., Suite 300, Nashville, TN 37203
Phone: 615-783-1300, Fax: 615-783-1301

Texas Rural Development State Office
Federal Building, Suite 102, 101 South Main, Temple, TX 76501
Phone: 254-298-1301, Fax: 254-298-1477

Utah Rural Development State Office
Federal Building, 125 State Street, Room 5428, Salt Lake City, UT 84138
Phone: 801-524-4063, Fax: 801-524-4406

Vermont Rural Development State Office
89 Main St., Montpelier, VT 05602
Phone: 802-828-6000, Fax: 802-828-6018

Virginia Rural Development State Office
Culpepper Building, Suite 238, 1606 Santa Rosa Rd., Richmond, VA 23229
Phone: 804-287-1550, Fax: 804-287-1718

Washington Rural Development State Office
1835 Black Lake Blvd., Suite B, Olympia, WA 98512-5715
Phone: 360-704-7742, Fax: 360-704-7742

West Virginia Rural Development State Office
75 High St., Room 320, Federal Building, Morgantown, WV 26505
Phone: 304-291-4793, Fax: 304-291-4032

Wisconsin Rural Development State Office
4949 Kirschling Court, Stevens Point, WI 54481
Phone: 715-345-7600, Fax: 715-345-7669

Wyoming Rural Development State Office
100 East B, Federal Building, Room 1005, Casper, WY 82602
Phone: 307-261-5271, Fax: 307-261-5167

CHAPTER 10 - STATE HELP FOR FARMERS AND RURAL RESIDENTS

Many state governments have followed the lead of the federal government to spur rural growth through agricultural, rural residency and business programs. New programs are being added on a continual basis. Check with your states finance authority or department of agriculture for new programs.

ALABAMA

Alabama Agricultural Development Authority
PO Box 3336, Montgomery, Alabama 36109-0336
(334)240-7245 Fax (334)240-7270
Beginning Farmer/Rancher Loan Program: Loans to beginning, qualifying Farmer/Rancher projects.
Alabama Swine Industry Development Project: Low interest loans (3%) for Swine Producers.

State Treasurer
Linked Deposits Office, 204 Alabama State House, Montgomery, AL 36130
The Wallace Plan For Linked Deposits Agricultural Loan: Low interest rate loans can be used for operating expenses and capital.
The Wallace Plan For Linked Deposits Small Business Loan: Low interest rate loans can be used for operating expenses and capital.

Alabama Department Of Agriculture and Industries
PO Box 3336, Montgomery, Alabama 36109, (334) 261-5872

ALASKA

Alaska Department Of Natural Resources
Division Of Agriculture, PO Box 949, Palmer, Alaska 99645
(907)745-7200 Fax (907)745-7112
Alaska Agriculture Loan Act: Six type of loans available to qualified farmers and ranchers that are residents of Alaska and can demonstrate experience in their agricultural field. Lower than conventional rate loans with varying re-payment terms. The six types of loans are: Farmer/Rancher development, construction and renovations of farm facilities, loans for livestock and equipment, short term operating loans, product processing and land clearing loans.
Wholesale Agricultural Market Development: Free technical assistance and counseling for market development.

Department of Community and Regional Affairs
Rural Development Fund, 333 W. 4th Ave., #220, Anchorage, AK 99501
(907)269-4500
Entrepreneur Rural Development Fund: $25,000 can be borrowed for construction, equipment, inventory and working capital to small businesses in rural areas.
Lender Participation Rural Development Fund: Matches up to $100,000 for large business projects as long as the business can raise an equal amount from other financing.

Community Enterprise Development Corporation
1577 C St., Suite 304, Anchorage, AK 99501
907-274-5400
Rural Development Loan Fund: Direct loans from $10,000 to $150,000 to be used by rural business for land, facilities, equipment, construction and expansion.

ARIZONA

No state agricultural programs available.

Arizona Department Of Agriculture
1688 W. Adams Street, Phoenix, Arizona 85007
(602)542-0978 Fax (602)542-5420

ARKANSAS

Arkansas Development Finance Authority
PO Box 8023, Little Rock, Arkansas 72203
(501)682-5939 Fax (501)682-5900
Beginning Farmer/Rancher Loan Program: Loans to beginning qualifying Farmer/ Rancher projects.
Capitol Access Program: Targets small business and agricultural related enterprises.
ADFA Bond Guaranty Program: For the development of the states industrial and agricultural value added processors, storage and distribution facilities and equipment.
Export Finance Program: Financial assistance to agribusiness exporting.
Farm Mediation: Assists in farmer/lender mediation of loan debt service problems.
Agricultural Cooperative Loan Program: Revolving loan fund to finance facilities and equipment used in crop diversification and assist limited resource farmers to make a living on small acreage.
Aquaculture Development Program: Serves the Aquaculture Industry.
Farm Link: Links retiring farmers/ranchers with potential buyers.

Arkansas State Plant Board
PO Box 1069, Little Rock, Arkansas 72203
(501)225-1598

CALIFORNIA

California Coastal Rural Development Corporation
PO Box 479, Salinas, California 93902
(408)424-1099 Fax (408)424-1094
Beginning Farmer/Rancher Programs: Funds must be used for crop production, harvest, farm ownership, improvements or purchasing equipment.

California Department Of Food And Agriculture
1220 N Street, Room 274, Sacramento, California 95814

COLORADO

Colorado Agricultural Development Authority
700 Kipling, Suite 4000, Lakewood, Colorado 80215
(303)239-4114 Fax (303)239-4125
Beginning Farmer/Rancher Loan Program: Assists first time farmers and ranchers in purchasing farmland and equipment.
Quality Agricultural Loan Program: Provides fixed rate financing on working capitol.
Colorado Capital Reserve/Ag: Assists banks in making loans which they would not ordinarily approve.

Department Of Agriculture
700 Kipling, Suite 4000, Lakewood, Colorado 80215
(303)239-4114

CONNECTICUT

No state agricultural programs available.

Connecticut Department Of Agriculture
165 Capitol Ave., Room 273, Hartford, Connecticut 06106
(860)566-3671 Fax (860)566-6576

DELAWARE

No state agricultural programs available.

Delaware Department Of Agriculture
2320 S. Dupont Highway, Dover, Delaware
(302)739-4811 Fax (302)573-6554

FLORIDA

No state agricultural programs available.

Florida Department Of Agriculture
Mayo Building, Room 421, Tallahassee, Florida 32399
(904) 488-4032 Fax (904) 488-7127

GEORGIA

Georgia Development Authority
Agricultural Loan Division, 2082 E. Exchange Place, Suite 102, Tucker, GA 30084
(404)414-3400 Fax (770)414-3407
Insured Farm Loans: Insured loan program for capital expenses

Georgia Department Of Agriculture
19 Martin Luther King Jr. Drive, Atlanta, Georgia 30334
(404) 656-3368

HAWAII

Hawaii Department Of Agriculture
Agricultural Loan Division. PO Box 22159, Honolulu, Hawaii 96823
(808) 973-9460 Fax (808)973-9455
Agricultural Loan Programs: Assists with farm acquisition, improvement, soil and water conservation, operations, emergency, new farmers/ranchers.
Aquaculture Loan Program: Targeted at aquacluturists that are unable to obtain financing elsewhere.

IDAHO

Idaho Department Of Agriculture
PO Box 790, Boise, Idaho 83701, (208)334-2227 Fax (208) 334-2879
Idaho Rural Rehabilitation Loan Program: Offers financial assistance to individuals and organizations whose projects will provide for rural economic development and cannot obtain financing elsewhere.

ILLINOIS

Illinois Farm Development Authority
427 East Monroe, Suite 201, Springfield, IL 62701
(217)782-5792 Fax (217)782-3989
Beginning Farmer/Rancher Bond Program: Assistance for first time farmland buyers with a net worth less than $250,000.
State Guarantee Program For Restructuring Agricultural Debt: Assistance for existing farm operations that need to restructure or refinance debt..
Young Farmer/Rancher Guarantee Program: Assistance for young farmers purchasing capital assets.
Specialized Livestock Guarantee: Assists farmers who are acquiring, constructing or remodeling specialized livestock facilities.

Illinois Development Finance Authority
2 North LaSalle St., Suite 980, Chicago, IL 60602
(312)793-5586
Rural Development Loan: Loans of up to $150,000 for businesses in communities of a population of less than 25,000.

Illinois Department Of Agriculture
PO Box 19281, State Fairgrounds, Springfield, IL 62794
(217)782-6675

INDIANA

Indiana Development Finance Authority
One North Capitol Ave., Suite 320, Indianapolis, Indiana 46204
(317)233-4332 Fax (317)233-6786
Agricultural Loan and Rural Development Project Guaranty Program: Value added agricultural projects.

Indiana Department of Commerce
1 North Capitol, Suite 700, Indianapolis, IN 46204-2288
(317) 232-8888
Rural Development Program: Assists in the development of farmer to consumer sales such as farmers markets.

Indiana Office Of The Commissioner Of Agriculture
ISTA Center, 150 W. Market St., Suite 414, Indianapolis, Indiana 46204
(317) 232-8774

IOWA

Iowa Agricultural Development Authority
505 Fifth Ave., Suite 327, Des Moines, IA 50309
(515) 281-6444 Fax (515)281-8618
Beginning Farmer/Rancher Program: The oldest and largest state agricultural program for beginning farmers and ranchers.
Loan Participation Program: Assists low income and beginning farmers.

Iowa Department of Agriculture and Land Stewardship
Wallace State Office Building, East 9th and Grand, Des Moines, Iowa 50319
(515)281-5993 Fax (515)242-5015

KANSAS

Kansas Development Finance Authority
700 SW Jackson, Suite 1000, Topeka, Kansas 66603
(913) 296-6747 Fax (913)296-6810
Beginning Farmer/Rancher Loan Program: Financial assistance for first timer farmer/ranchers with land, improvements, equipment and breeding stock.

Kansas State Board Of Agriculture
901 S. Kansas Ave., Topeka, Kansas 66612
(913) 296-3736

KENTUCKY

Kentucky Department Of Agriculture
500 Mero St., 7th Floor, Frankfort, Kentucky 40601
(502) 564-4696 Fax (502) 564-6527
Investment Deposit Link: Assists farming operations and agribusiness's.

Cabinet for Economic Development
Office of Business and Technology, Capital Plaza Tower, 500 Mero St., Frankfort, KY 40601
(502)564-7670
Kentucky Rural Economic Development Authority Incentives: Assists businesses in establishing new and existing manufacturing operations in qualifying counties. Approved projects can be financed with conventional funds or by state bonds. Companies receive credits against state income tax.

LOUISIANA

Louisiana Agriculture Finance Authority
PO Box 3334, Baton Rouge, Louisiana 70821
(504) 922-1292 Fax (504)922-1289
Agribusiness Loan Guarantee Program: Assists ag producers and processors.

Louisiana Department Of Agriculture and Forestry
PO Box 3334, Baton Rouge, Louisiana 70821
(504)922-1277

MAINE

Maine Finance Authority
83 Western Ave., PO Box 949, Augusta, ME 04332
(207)623-3263
The Natural Resource Entrants Loan Insurance Program: Targets Maine Farmers who meet certain eligibility criteria.

Maine Department Of Agriculture
State House Station No. 28, Augusta, Maine 04333
(207)287-3177

MARYLAND

Maryland Department Of Agriculture
40 Harry S. Truman Parkway, Annapolis, MD 21401
(410)841-5855
Maryland Agricultural Water Quality Cost Share: Provides money to individuals or businesses to install practices that protect water quality.

MASSACHUSETTS

No state agricultural program available.

Massachusetts Department Of Food And Agriculture
100 Cambridge St., Boston, MA 02202
(617)727-3018

MICHIGAN

Michigan Department Of Agriculture
611 W. Ottawa, PO Box 30017, Lansing, Michigan 48909
(517) 335-3401
Beginning Farmer/Rancher Program: Assists beginning farmer/ranchers and ag manufacturers.

MINNESOTA

Minnesota Rural Finance Authority
90 West Plato Blvd. , St. Paul, Minnesota 55107
(612)297-3557
The Basic Farm Loan Program: Assist qualified beginning farmers/ranchers with the purchase of real-estate.
Agricultural Improvement Loan Program: Provides financing for low equity farmers for items such as facilities, storage, wells and erosion control.
Seller Assisted Loan Program: Assists seller of ranch or farm willing to fund a portion of property to beginning farmers/ranchers.
Livestock Expansion Loan Program: Creates affordable financing for livestock farmers for new improvements to buildings or structures used for livestock production.
Restructure II Loan Program: Assists with reorganization of farm/ranch debt.
Agricultural Development Bond Program: Beginning Farmer/Rancher Program.

Minnesota Department Of Agriculture
90 W. Plato Blvd., St. Paul, Minnesota 55107
(612)297-2301

MISSISSIPPI

Mississippi Business Finance Corporation
PO Box 849, Jackson, Mississippi 39205
(601)359-3552
Agribusiness Loan Program: Assists agribusiness in the state of Mississippi
Emerging Crop Program: Loans to qualified borrowers to finance non-land capital expenses to establish production of any non-traditional plant or animal crop for which there is a growing public demand.

Mississippi Department Of Agriculture and Commerce
PO Box 1609, Jackson, Mississippi 39215
(601)354-7098

MISSOURI

Missouri Agricultural and Small Business Development Authority
PO Box 630, Jefferson City, MO 65102
(314) 751-2129
Beginning Farmer/Rancher Program: Assists beginning qualified farmer/ranchers.
Single Purpose Animal Facilities Loan Guarantee Program: Loans for single purpose livestock facilities or to expand current facilities.

Missouri Department Of Agriculture
PO Box 630, Jefferson City, MO 65102
(314)751-2613

MONTANA

Montana Department Of Agriculture
Agriculture/Livestock Building, Capitol Station, Helena, MT 59620
(406)444-2402
Rural Assistance Loan Program: Available to Farmers and Ranchers unable to qualify for conventional financing. Funds can be used for operation and land.
FSA Subordination Loan Program: Provides annual operating funds for FSA Borrowers.
Junior Agriculture Loan Program: Loans for 4-H and FFA youth.

NEBRASKA

Nebraska Investment Finance Authority
1230 "O" Street, Suite 200, Lincoln, Nebraska 68508
(402)434-3900
IDB Based Agricultural Loan Program: Beginning Farmer/Rancher Program

Nebraska Department Of Agriculture
PO Box 97947, Lincoln, Nebraska 68508

NEVADA

Nevada Division Of Agriculture
350 Capitol Hill Ave., Reno, Nevada 89510
(702) 688-1180
Junior Agriculture Loan Program: Assists youth 9 to 21 years old and can be used to purchase agricultural inputs or breeding livestock.

NEW HAMPSHIRE

No state agricultural programs available

Department Of Agriculture, Markets and Food
PO Box 2042, Concord, New Hampshire 03302-2042
(603) 271-3551

NEW JERSEY

Department Of Agriculture
John Fitch Plaza, CN 330, Trenton, New Jersey 086625
(609) 984-2506
Agricultural Economic Investment Opportunity Loan Program: Loan to businesses involved in producing agricultural commodities.

NEW MEXICO

No state agricultural programs available.

New Mexico Department Of Agriculture
PO Box 30005, Las Cruces, NM 88003
(505)646-2804 Fax (505)646-2186

NEW YORK

No state agricultural programs available.

New York State Department Of Agriculture and Markets
1 Winners Circle, Capitol Plaza, Albany, New York 12235
(518)457-2771 Fax (518)457-3087

New York Job Development Authority
605 3rd Ave., New York, NY 10158
(212)818-1700
Rural Development Loan Fund: Lost cost financing for small businesses in distressed rural areas of less than 25,000.

Rural Areas Development Fund: Provides low cost financing for businesses and public entities that provide jobs for displaced farm families or to supplement farm family income.

NORTH CAROLINA

NC Agricultural Finance Authority
PO Box 27908, Raleigh, NC 27611-7908
(919)733-0635 Fax (919)733-1460
Beginning Farmer Loan Program: Assists the first time farmer/rancher.
Series I Farm Real Estate Loans: Loans to farmers and ranchers having trouble qualifying for conventional financing.
Agricultural Development Bonds: Bonds issued for the processing and manufacturing of agricultural products.

North Carolina Department Of Agriculture
PO Box 27647, Raleigh, North Carolina 27611

NORTH DAKOTA

The Bank Of North Dakota
PO Box 5509, Bismark, North Dakota
(701) 328-5672 Fax (701) 328-5832
North Dakota Agri-Bond Program: Tax exempt bonds issued for manufacturing and processing agricultural products.
Family Farm Loan Program: Loans for Farmers with new worth less than $150,000 to be used for equipment, livestock or refinancing real estate.
AG Pace Program: Businesses that can supplement farm income.
Beginning Farmer/Rancher Real Estate Loan Program: Qualified beginning Farmer/ Ranchers purchasing land.

North Dakota Department Of Agriculture
600 E. Blvd., Bismark, North Dakota 58505
(701) 328-2231

OHIO

Ohio Agricultural Financing Commission
65 S. Front St., Room 606, Columbus, Ohio 43215
(614) 466-2732 Fax (614)466-6124
Agricultural Financing Bonds: Help for beginning Farmers/Ranchers that have not had any direct or indirect interest in substantial farmland.

Ohio Department Of Agriculture
65 S. Front St., Columbus, Ohio 43215
(614)752-9815 Fax (614) 644-5017

OKLAHOMA

Oklahoma Development Finance Authority
301 NW 63rd Street, Suite 225, Oklahoma City, OK 73112
(405)842-1145 Fax (405)848-3314
Beginning Farmer Loan Program: Loans to first time qualified Farmers/Ranchers.
Agriculture Linked Deposit Program: Loans to alternative enterprise or agribusiness that cannot find financing through a conventional lender.

Rural Enterprises
422 Cessena Drive, Durant, OK 74701
(405)924-5094
Assists with financial packages and enterprise planning in rural areas.

Oklahoma Department Of Agriculture
2800 N. Lincoln Blvd., Oklahoma City, Oklahoma 73105
(405) 521-3863 Fax (405) 521-4912

OREGON

No state agricultural programs available.

Oregon Department Of Agriculture
635 Capitol Street NE, Salem, Oregon 97310-011
(503)378-4152 Fax (503)373-1947

PENNSYLVANIA

No state agricultural programs available at this time.

Pennsylvania Department Of Agriculture
2301 North Cameron Street, Harrisburg, PA 17110-9408
(717)787-8460 Fax (717)787-1858

RHODE ISLAND

No state agricultural programs available at this time.

Rhode Island Department Of Agriculture
22 Hoyes Street, Providence, Rhode Island 02908
(401) 277-2781 Fax (401)277-6047

SOUTH CAROLINA

No state agricultural programs available.

South Carolina Department Of Agriculture
Wade Hampton Office Building, PO Box 11280, Columbia, South Carolina 29211-1280
(803)734-2210 Fax (803)734-2192

South Carolina Jobs/Economic Development Authority
1201 Main Street, Suite 1750, Columbia, SC 29201
(803)737-0079
Rural Development Administration Intermediary Relending Program: Program is intended to provide supplemental or gap financing for small business in areas of 25,000 or less. Loans can be up to $150,000 and must be repaid within 7 years.

SOUTH DAKOTA

South Dakota Department Of Agriculture
523 East Capitol Ave., Pierre, South Dakota 57501-3182
(605)773-5436 Fax (605)773-3481
Beginning Farmer Program: Assists young, low equity, beginning Farmers/Ranchers.
Value Added Livestock Underwriting Program: Money for qualified Farmers/Ranchers to obtain livestock.
Livestock Loan Participation Program: Money for qualified Farmers/Ranchers to obtain livestock.

TENNESSEE

Tennessee Department Of Agriculture
PO Box 40627, Nashville, Tennessee 37204
(615)360-0160 Fax (615)360-0194
Beginning Farmer Program: Assists young, low equity, beginning Farmers/Ranchers.

TEXAS

Texas Agricultural Finance Authority
PO Box 12847, Austin, Texas 78711
(512)463-7686 Fax (512)475-1762
Young Farmer Loan Guarantee Program: Assists beginning Farmers/Ranchers.
Loan Guaranty Program: Financing for agricultural endeavors related to innovative, value added or diversified products.
Farm And Ranch Finance Program: Assists qualified Farmers and Ranchers with land purchases.
Rural Microenterprise Loan Program: Loans to family owned enterprises in rural Texas.

Texas Department Of Agriculture
PO Box 12847, Capitol Station, Austin, Texas 78711
(512)463-7462 Fax (512)463-9968

UTAH

Utah Department Of Agriculture
350 North Redwood Road, PO Box 146500, Salt Lake City, Utah 84114
(801)538-7176 Fax (801)538-7126
Agriculture Resource Development Loan Program: Low interest loans to Farmers and Ranchers for projects that meet conservation and pollution goals.
Rural Rehabilitation Loan Program: Low interest loans for qualified Farmers and Ranchers for real estate, operations, education and irrigation projects.

VERMONT

Vermont Industrial Development Authority
56 East State Street, Montpelier, VT 05602
(802)223-7226
Agricultural Finance Program: Low interests loans to family farmers for real estate and equipment.

Vermont Department Of Agriculture
116 State Street, Drawer 20, Montpelier, Vermont 05620-2901
(802)832-2430 Fax (802)832-2361

VIRGINIA

Virginia Department Of Agriculture
1100 Bank Street, Richmond, Virginia 23219
(804)786-3538 Fax (804)371-7679
Beginning Farmer Program: Assists young, low equity, beginning Farmers/Ranchers.

WASHINGTON

No state agricultural programs available at this time.

Washington Department Of Agriculture
PO Box 42560, Olympia, Washington 98504-2560
(360)902-1800 Fax (360)902-2092

WEST VIRGINIA

West Virginia Department Of Agriculture
State Capitol, Room E-28, Charleston, West Virginia 25305
(304)558-2201 Fax (304)558-0451
Rural Rehabilitation Loan Fund: Farmers, Ranchers, 4-H, FFA Members and any Agribusiness's can apply for low interest loans.

WISCONSIN

Wisconsin Department Of Agriculture
2811 Agriculture Drive, Madison, Wisconsin 53708
(608)224-5053 Fax (608)224-5110
Beginning Farmer Bond Program: Offers beginning Farmers and Ranchers reduced interest rates for capital purchases.
Agriculture Development and Diversification: Grants provided to Farmers, Ranchers and Cooperatives to develop new diversified products and markets.
Credit Relief Outreach Program: Assists Farmers and Ranchers who are unable to obtain conventional financing.
Farm Asset Reinvestment Management Program: Assists Farmers and Ranchers seeking to expand their operations.
Agribusiness Fund: Loans for agribusiness's developing a new product or improving processing methods.

Department of Development
123 W. Washington Ave., Madison, WI 53707
(608)266-3278
Rural Economic Development Programs: Grants and loans for planning and managing a small business in rural areas.

WYOMING

Wyoming State Land and Farm Loan Office
122 West 25th St., Herschler Building 3W, Cheyenne, Wyoming 82002
(307)777-7332 Fax (307)777-5400
State Farm Loan Board: Low interest loans for Farmers and Ranchers.

Glossary

ACREAGE ALLOTMENT. The individual farm's share, based on previous production, of the national acreage needed to produce sufficient supplies of specific crops like peanuts, rice, cotton and tobacco.

ACREAGE BASE. The individual farm's program average acreage for a commodity planted for harvest during the previous five years. The acreage base is used in acreage reduction and paid land diversion programs.

ACREAGE REDUCTION PROGRAM (ARP). A voluntary land retirement program in which farmers idle a prescribed portion of their crop acreage base of wheat, feed grains, cotton or rice. Farmer must participate to be eligible for Commodity Credit Corporation (CCC) loans and deficiency payments. The Secretary of Agriculture sets the ARP each fall for the coming year.

AGRONOMY. The science of crop production and soil management.

ALTERNATIVE AGRICULTURE. Production methods other than energy and chemical intensive on crop farming. Alternatives include using animal and green manure rather than chemical fertilizers, integrated pest management instead of chemical pesticides, reduced tillage, crop rotation, alternative crops, or diversification of the farm enterprise.

ARTIFICIAL INSEMINATION (AI). The mechanical injection of semen into the womb of the female animal with a syringe like apparatus.

AQUACULTURE. The production of aquatic plants or animals in a controlled environment.

BASIC COMMODITIES. Six crops (corn, cotton, peanuts, rice, tobacco and wheat) that are covered by price support programs.

BILATERAL AGREEMENTS. Trade pacts or general concessions entered into between two nations.

BIOTECHNOLOGY. The use of technology, based on living systems, to develop products for commercial and other purposes. Examples include plant regeneration and gene manipulation and transfer.

CARRYOVER. Existing supplies of farm commodity not used at the end of a marketing year, and remaining to be carried over to the next year. Marketing years generally start at the beginning of a new harvest for commodity and extend to the same time in the following year.

CASH GRAIN FARM. A farm on which corn, grain, sorghum, small grains, soybeans or field beans and peas account for at least 50 percent of the value of products sold.

CENSUS OF AGRICULTURE. A count taken every five years of the number of farms, land in farms, crop acreage, production information, farm value and farm products, etc.

CHECKOFF PROGRAMS. Producers and importers pay a given amount per unit of production marketed in order to finance market development or research programs.

COMMODITY CREDIT CORPORATION (CCC). A federally owned corporation managed by the Department of Agriculture. It was created to stabilize, support and protect farm income and prices through loans, purchases, payment and other programs.

COMBINE. A self propelled machine for harvesting grain and other seed crops. In one operation it cuts, threshes, separates, and cleans the grain and scatters the straw.

CONSERVATION PLAN. A combination of land uses and practices to protect and improve soil productivity and to prevent soil deterioration.

CONSERVATION PRACTICES. Methods which reduce soil erosion and retain soil moisture. Major conservation practices include conservation tillage, crop rotation, contour farming, strip cropping, terraces, diversions and grassed waterways.

CONSERVATION RESERVE PROGRAM (CRP). Lands deemed redouble can be put under a 10-year federal conservation set aside program. Land owners received annual rental payments in exchange for agreeing to take land from production. The federal government determines which bids are accepted.

CONSERVATION TILLAGE. Any of several farming methods that provide for seed germination, plant growth, and weed control, yet maintain effective ground cover throughout the year and disturb the soil as little as possible. The aim is to reduce soil loss and energy use while maintaining crop yields and quality.

CONTOUR FARMING. Field operations such as plowing, planting, cultivating, and harvesting on the contour, or at right angles to the natural slope to reduce soil erosion, protect soil fertility and use water more effectively.

COOPERATIVE. An organization formed for the purpose of producing and marketing goods or products owned collectively by members how share in the benefits.

COOPERATIVE EXTENSION SYSTEM. A system of State, local and Federal organization working together to provide practical educational services outside the classroom on agriculture, household management, and other topics. States participate mostly through their Land Grant Universities, while the Federal partner is the USDA's Extension Service.

CROP ROTATION. The practice of growing different crops in succession on the same land.

CUSTOM WORK. Specific farm operations performed under contract between the farmer and the contractor. The contractor furnishes labor, equipment and materials to perform the operation. Custom harvesting of grain, spraying and picking of fruit and sheep shearing are examples of custom work.

DEFICIENCY PAYMENTS. Funds paid by the Commodity Credit Corporation to farmers for certain commodities when farm prices are below a target price. It is based on the difference between the price level established by law (target price) and the higher of the price support or loan rate and the market price. Generally, the federal government pays this difference to a farmer who qualifies (by meeting all farm program conditions) for that portion of the farmer's production specified in the farm program.

DISASTER PAYMENTS. Federal payments made to farmers because of a natural disaster when planting is prevented or crop yields are abnormally low because of adverse weather and related conditions. Disaster payments may be provided under existing legislation or under special legislation enacted after an extensive natural disaster.

DOUBLE CROP. Two different crops grown on the same area in one growing season.

DRYLAND FARM. A system of producing crops in semi arid regions (usually with less than 20 Inches of annual rainfall) without the use of irrigation.

EROSION. The process in which water or wind moves soil from one location to another.

ETHANOL. An alcohol fuel that may be produced from an agricultural foodstock such as corn, sugarcane or wood. It is blended with gasoline to enhance octane and reduce automobile exhaust pollution.

EXPORT ENHANCEMENT PROGRAM (EEP). A program inaugurated in 1985 to offset subsidies offered by other countries. The program is extended to third-party countries who would otherwise buy from the subsidizing countries.

EXPORT SUBSIDIES. Special incentives, such as cash payments, tax exemptions, preferential exchange rates, and special contracts extended by governments to encourage increased foreign sales.

FAMILY FARM. An agricultural business which produces agricultural commodities for sale in such quantities so as to be recognized as a farm rather than a rural residence; produces enough income to pay family farm operating expenses, to pay debts and to maintain the property; is managed by the operator; has a substantial amount of labor provided by the operator and family; may use seasonal labor during peak periods and a reasonable amount of full time hired labor.

FARM BILL. The omnibus agricultural legislation that expires every 4 or 5 years.

FEDERAL INSECTICIDE, FUNGICIDE & RODENTICIDE ACT (FIFRA). The federal law governing the registration and use of agricultural chemicals.

FEED GRAIN. Any of several grains most commonly used for livestock or poultry feed, such as corn, sorghum, oats, rye and barley.

FERTILIZER. Any organic or inorganic material which is added to soil to provide nutrients for plant growth.

FLOOD PLAINS. Lowland and relatively flat areas adjoining inland and coastal waters including floodprone areas of islands. This land includes, at a minimum, those areas that are subject to a 1% or greater chance of flooding in any given year.

FORAGE. Vegetable matter, fresh or preserved, that is gathered and fed to animals as roughage; includes alfalfa, hay, corn silage, and other hay crops.

FORWARD CONTRACTING. A method of selling crops before harvest by which the buyer agrees to pay a specified price to a grower for a portion, or all of the growers crops.

FUTURES CONTRACT. An agreement between two people, one who sells and agrees to deliver and one who buys and agrees to receive a certain kind, quality and quantity of a product during a specific delivery month.

GENETIC ENGINEERING. Genetic modification of organism by recombinant DNA, recombinant RNA, or other specific molecular gene transfer or exchange techniques.

GENERAL AGREEMENT ON TARIFFS AND TRADE (GATT). An International pact that governs much of the world trade. The agreement consists of rules of conduct in international trade, a system for resolving trade disputes and a forum for conducting multi-lateral trade discussion.

GLEANING. Collecting of unharvested crops from the fields, or obtaining agricultural products from farmers, processors or retailers without charge.

GREAT PLAINS. A level of gently sloping region of the United States that lies between the Rockies and approximately the 98th meridian. The area is subject to recurring droughts and high winds. Consists of parts of North Dakota, South Dakota, Montana, Nebraska, Wyoming, Kansas, Colorado, Oklahoma, Texas and New Mexico.

GREEN MANURE. Any crop or plant grown and plowed under to improve the soil by adding organic matter and subsequently releasing plant nutrients, especially nitrogen.

GROSS FARM INCOME. Income which farm operators realize from farming. It includes cash receipts from the sale of farm products, government payments, value of food and fuel produced and consumed on farms, and the rental value of farm dwellings.

GROUND WATER. Water beneath the Earth's surface between saturated soil and rock, which supplies wells and springs.

HERBICIDE. Any chemical used to destroy plants, especially weeds.

HYDROPONICS. Growing of plants in water containing dissolved nutrients, rather than in soil. This process is being used in green houses for intensive off-season production of vegetables.

INTEGRATED PEST MANAGEMENT (IPM). An integrated approach to controlling plant pests using careful monitoring of pests and weeds. It may include use of natural predators, chemical agents and crop rotations.

LAND GRANT UNIVERSITIES. Institutions including State colleges and universities and Tuskegee University, eligible to receive funds under the Morrill Acts of 1862 and 1890. The federal government granted land to each State and territory to encourage practical education in agriculture, homemaking and mechanical arts.

LOAN DEFICIENCY PAYMENTS. Commodity Credit Corporation payments provided to producers who, although eligible to obtain a marketing loan for wheat, feed grains, upland cotton, rice, or oilseed crop agree to forgo obtaining the loan. The payment is determined by multiplying the loan payment rate by the amount of commodity eligible for loan. The payment rate per unit is the announced loan level minus the repayment level used in the marketing loan.

LOAN RATE. The price per unit (bushel, bale, pound) at which the government will provide loans to farmers to enable them to hold their crops for later sale.

LOW INPUT SUSTAINABLE AGRICULTURE (LISA). Alternative methods of farming that reduce the application of purchased inputs such as fertilizer, pesticides, and herbicides. The goals of these alternative practices are to diminish environmental hazards while maintaining or increasing farm profits and productivity. Methods include crop rotations and mechanical cultivation's to control weeds; integrated pest management strategies such as introducing harmless natural enemies; planting legumes that transform nitrogen from the air into a form plants can use; application of livestock manures and compost for fertilizer; and overseeding of legumes into maturing fields of grain crops, or as post season cover crops to curtail soil erosion.

MARKETING SPREAD. The difference between the retail price of a product and the farm value of the ingredients in the product.

MOST-FAVORED NATION (MFN). Agreements between countries to extend the same trading privileges to each other that they extend to any other country.

NET FARM INCOME. The money and non-money income farm operators realized from farming as a return for labor, investment and management, after production expenses have been paid.

NONFARM INCOME. Includes all income from nonfarm sources (excluding money earned from working for other farmers) received by farm operator households.

NONRECOURSE LOANS. Price-support loans to farmers that enable them to hold their crops for later sale. Farmers may redeem their loans by paying them off with interest. The loans are "non-recourse" because if a farmer cannot profitably sell the commodity and repay the loan when it matures, the commodity on which the loan was advanced can be delivered to the government for settlement of the loan.

NON-TARIFF BARRIER. Health or sanitary standards established by countries which tend to restrict the flow of trade between nations.

OFF-FARM INCOME. Includes wages and salaries from working for other farmers, plus non-farm income, for all owner operator families (whether they live on the farm or not).

ORGANIC FARMING. A production system that completely or mostly excludes the use of synthetically compounded fertilizers, pesticides, or growth regulators.

PAID LAND DIVERSION. Land idled under a farm program for which a payment is made. The level of payment is established in advance and farmers participate on a voluntary basis. The paid diversion is allowed only after other non-paid acreage diversion requirements are met.

PAYMENT LIMITATIONS. Limitations set by law on the amount of money any one person may receive in farm program payments each year under the feed grain, wheat, cotton, and rice programs.

PERCOLATION. The downward movement of water though soil under the influence of gravity.

PESTICIDE RESIDUE. A detectable level of chemical residue found on a food product.

PESTICIDE TOLERANCE LEVELS. Scientifically acceptable level of a pesticide residue that can exist on a fruit or vegetable product. Usually expressed in parts per million or billion.

PRICE INDEX. An indicator of average price change for a group of commodities that compares price for those same commodities in some other period, commonly called the base period.

PRICE SUPPORT LEVEL. The price for a unit of a farm commodity (bushel, pound) which the government will support through price support loans and/or payments. Price support loans and/or payments. Price support levels are determined by law and are set by the Secretary of Agriculture.

PRICE SUPPORT PROGRAMS. Government programs that aim to keep farm prices received by participating producers from falling below specific minimum levels. Price support programs form major commodities are carried out by providing loans and purchase agreements to farmers so that they can store their crops during periods of low prices. The loans can later be redeemed if commodity prices rise sufficiently to make the sale of the commodity on the market profitably or the farm can forfeit the commodity to the Commodity Credit Corporation. With a purchase agreement, the producer may sell the commodity to the CCC.

RANGELAND. Land which is predominantly grasses, grasslike plants, or shrubs suitable for grazing.

RENEWABLE RESOURCES. Resources such as forests, rangeland, soil and water that can be restored and improved.

RUMINANT. Animals having a stomach with four compartments. Their digestive process is more complex that that of animals having a true stomach. Ruminants include cattle, sheep, goats, deer, bison, buffalo, camels and giraffes.

RURAL. An area that has a population of fewer than 2,500 inhabitants and is outside an urban area. A rural area does not apply only to farm residences or to sparsely settled areas, since a small town is rural as long as it meets the above criteria.

SET ASIDE. A government farm program term used to describe the acreage a farmer must devote to soil conserving uses (such as grasses, legumes, and small grain which is not allowed to mature) in order to be eligible for production adjustment payments and price- support loans.

SILAGE. Prepared by chopping green forage (grass, legumes, field corn, etc.) into an air tight chamber, where it is compressed to exclude air and undergoes an acid fermentation that retards spoilage. Contains about 65% moisture.

SILVICULTURE. A branch of forestry dealing with the development and care of timber.

SODBUSTER. A determination made under the conservation compliance program as to which soils are considered "erodible" and may not be tilled.

STATE AGRICULTURAL EXPERIMENT STATION. State operated institutions, established the Hatch Act of 1887 and connected to land grant universities in each State, which carry outreach of local and regional importance in the areas of food, agriculture and natural resources.

SUBSISTENCE FARM. A low income farm where the emphasis is on production for use of the operator and the operators family rather than for sale.

SUSTAINABLE AGRICULTURE. An integrated system of farming that will, over the long term, satisfy food and fiber needs, enhance environmental quality, make the most efficient use of resources, sustain the economic viability of farm operations and enhance the quality of life.

SWAMPBUSTER. A determination made under the conservation compliance program as to which soils are considered "wetlands" and may not be tilled.

"T" VALUES. A local soil conservation district determination of what is an acceptable rate of erosion on a given soil.

TARGET PRICES. A minimum level of prices determined by law to provide an economic safety net. Sometimes called the "guaranteed price level." The target price, based on costs of production, becomes the price support level at which the government will bolster farm income by making price support payments to qualifying farmers when national average market prices fall below the target. See DEFICIENCY PAYMENTS.

TERMINAL MARKET. A metropolitan market that handles all agricultural commodities.

UNIT COST. The average cost to produce a single item. The total cost divided by the number of items produced.

URBAN. A concept defining an area that has a population of 2,500 or more inhabitants.

WATERSHED. The total land area, regardless of size, above a given point on a waterway that contributes run off water to the flow at that point. A major subdivision of a drainage basin. The United States is generally divided into 18 major drainage areas and 160 principal river drainage basin containing some 12,700 smaller waterways.

WETLANDS. Land that is characterized by an abundance of moisture and that is inundated by surface or ground water often enough to support a prevalence of vegetation typically adapted for life in saturated soil conditions.

WHOLESALE PRICE INDEX. Measures of average changes in prices of commodities in primary U.S. markets. "Wholesale" refers to sales in large quantities by producers, not in prices received by wholesalers, jobbers, or distributors. In agriculture it is the average price received by farmers for their farm commodities at the first point of sale when the commodity leaves the farm.

ZOONOTIC DESEASES. Diseases that, under natural conditions, are communicable from animals to humans.

4-H. Club for young people (9-19 years old) sponsored by the Agricultural Extension Service to foster agricultural, homemaking and other skills.

Suggested Reading

How To Make $100,000 Farming 25 Acres, Booker T. Whatley
The American Botanist, PO Box 532, Chillicothe, IL 615213

Successful Small-Scale Farming: An Organic Approach, Karl Schwenke
Storey Communications, Pownal, Vermont 05261

Back To Basics, Reader's Digest
Reader's Digest Association, Pleasantville, NY

Moving to a Small Town, Wanda Urbanska and Frank Levering
Fireside Books, New York, NY

How to Find Your Ideal Country Home, Gene GeRue
Heartwood Publications, Zanoi, MO 65784

Country Dreams, Alan Schabilion
Misty Mountain Press, Little Switzerland, NC

Five Acres and Independence, M.G. Kains
Dover Publications, New York

How to Buy and Enjoy a Small Farm, George Laycock
McKay, New York

Getting Started In Farming Series:

Getting Started in Farming; Mostly on Your Own Part-Time or Small Farms
So You Have Inherited A Farm
Small is Bountiful
Getting Started in Farming; On A Small Scale
Overview of Small Farm Programs at the Land Grant Colleges and Universities
Directory of State Extension Small Farm Contacts
Small Farm Digest
Proceedings of the National Small Farm Conference (1996)
Getting Help for Your Small Farm from the USDA
Brochure on Small Farm Program

To obtain any of the above: Small Farm Program, USDA-CSRESS, Stop 2220, Washington, DC 20250-2220, Phone: 1-800-583-3071

Acreage Advisor Newsletter
15400 North 56th, Lincoln, NE 68514

Alternative Agriculture News
9200 Edmonston Road, Suite 117, Greenbelt, MD 20770

Small Farm Today
Ridge Top Ranch, Rt 1, 3903 W. Ridge Trail Rd., Clark, MO 65243

The Small Farmer
Dept. 225, PO Box 1627, Sisters, OR 97759

Farm Times
707 F St.
Rupert, ID 83350